THE SLEEP SOLUTION

W. CHRIS WINTER is a neurologist and internationally recognised sleep medicine specialist, with 24 years of experience. He is a sought-after consultant for professional sports organisations, US military groups, and large corporations. Dr Winter is the *Men's Health* magazine sleep advisor, regularly blogs for *The Huffington Post*, and has written for a wide variety of magazines, including *Women's Health*, *Runner's World*, *Triathlete*, and *Details*.

THE
SLEEP
SOLUTION

Why Your Sleep Is Broken
and How to Fix It

DR W. CHRIS WINTER

SCRIBE

Melbourne • London

Scribe Publications
18–20 Edward St, Brunswick, Victoria 3056, Australia
2 John St, Clerkenwell, London, WC1N 2ES, United Kingdom

Published in Australia and the United Kingdom by Scribe 2017
Published by arrangement with New American Library, an imprint of Penguin Publishing Group, a division of Penguin Random House LLC

Every effort has been made to ensure that the information contained in this book is complete and accurate. However, neither the publisher nor the author is engaged in rendering professional advice or services to the individual reader. The ideas, procedures, and suggestions contained in this book are not intended as a substitute for consulting with your physician. All matters regarding your health require medical supervision. Neither the author nor the publisher shall be liable or responsible for any loss or damage allegedly arising from any information or suggestion in this book. The opinions expressed in this book represent the personal views of the author and not of the publisher.

While the author has made every effort to provide accurate telephone numbers, Internet addresses, and other contact information at the time of publication, neither the publisher nor the author assumes any responsibility for errors, or for changes that occur after publication. Further, publisher does not have any control over and does not assume any responsibility for author or third-party Web sites or their content.

Accordingly nothing in this book is intended as an express or implied warranty of the suitability or fitness of any product, service or design. The reader wishing to use a product, service or design discussed in this book should first consult a specialist or professional to ensure suitability and fitness for the reader's particular lifestyle and environmental needs.

Jacket design and illustration by Sandra Chiu
Book design by Pauline Neuwirth
Printed and bound in the UK by CPI Group (UK) Ltd, Croydon CR0 4YY

Scribe Publications is committed to the sustainable use of natural resources and the use of paper products made responsibly from those resources.

9781925322033 (Australian edition)
9781911344315 (UK edition)
9781925548174 (e-book)

CiP data records for this title are available from the National Library of Australia and the British Library.

scribepublications.com.au
scribepublications.co.uk

To my patients,

both to the ones I've tried to help and the ones I've yet to meet,

I humbly wrote this book for you.

To my wife, Ames,

you are my love and inspiration,

I solely wrote this book because of you.

Contents

Prologue

HAVE ALWAYS LOVED SLEEP, AND it's always been important to me. I remember as a child appreciating how fantastic it was to sleep in on the weekend. I have very clear memories of waking up for school as snow fell and eagerly scanning the radio to listen for school closing information. Finding out schools were closed meant getting right back into bed for extra sleep! Because both of my parents were public school teachers, it was always a family event.

When I was seven years old, my doctor prescribed me medicine for a bad cold. It had to be given around the clock, so at some point during the night, my mother woke me up to swallow some strongly flavored antibiotic liquid. The nocturnal awakening and subsequent sleep always seemed to make the night feel longer. I loved it.

I decided to become a doctor in third grade because I liked drawing organs and memorizing the Latin names for muscles. Family members and friends always gave me such high praise when I told them my plans, so I'm sure that solidified my goal even further. As time passed, I would go through dermatology phases, pediatric phases, and even orthopedic phases, but life decisions and luck eventually landed me in the field of sleep.

I started learning about sleep and studying it long before I became a doctor, even before medical school. I was fascinated by the study of sleep, running sleep studies and getting my hands dirty when it came to research. And get dirty they did. My hands were plenty dirty studying sleep apnea in Yucatán micro pigs when I was an undergraduate. Pigs happen to be a fantastic model for sleep, and they can snore just as loudly as any human sleep apnea patient. For those who are not familiar with the Yucatán micro pig, there is little "micro" about them, except for their patience as a teenager tries to shave their little tail and tape a probe to it. When it came to sleep, smelling like a farm was a small price to pay.

My curiosity has continued to be unusually strong. As a physician, I like to know as much as I can about what my patients are going through. To that end, over the years, I've volunteered to have my blood drawn, and I've undergone a three-hour neuropsychological test battery. I've had a nasogastric tube stuck in my nose, my muscles electrocuted, lidocaine injected into my love handle, making it go numb. I even had a powerful electromagnet applied to my head, causing my arm to spasm uncontrollably.

My medical experimentation reached its peak when during a boring medical call-night I asked if I could jump into an MRI scanner to take some pictures of my brain so I could see what the experience

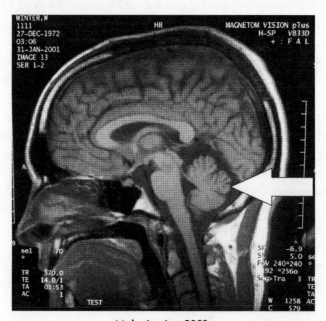

My brain, circa 2001.

was like and what was going on up there. All of my patients said it was loud, claustrophobia inducing, and generally miserable. I was fairly unimpressed. What did impress me was the size of my cerebellum—strangely small. I posted my MRI in the neurology residents' reading room the next morning. It was a tradition to post unusual images or diagnostic dilemmas so other residents could write their guesses and theories next to the images. For those who didn't notice my name on the films, virtually everyone wrote "cerebellar hypertrophy," or an unusually small cerebellum. Unexpectedly, my cerebellum (the part of the brain responsible for muscle coordination; indicated by an arrow in the photo) was a little puny, as you can see from the actual image. Of those who did notice my name, the overwhelming guess was "testicular atrophy." Smart-asses.

The bottom line is this: Despite some occasional unwelcome information, I like experiencing what my patients experience. It engenders trust and a common ground from which to work. I want to help my patients with their issues and understand what they are going through as much as I can.

As a sleep specialist, I help patients with their sleep problems every day. I'm also lucky enough to work with many professional athletes and help solve their sleep problems too. This might mean helping a team plan out the best time to travel during a long road trip. It might mean helping an athlete and his family adjust to a new baby in the house. Often athletes experience difficulty sleeping before big games or after poor performances. Regardless of the situation, I hope to help players improve their performance by improving their sleep.

The great thing about sleep is that it crosses so many groups of individuals. Over the years, I have had the fortune of working with elite members of the United States military and technology corporations as well as students across the country, helping them achieve better performance through improved sleep. These experiences have made me a better physician to my patients.

It is a rewarding occupation. From my desire to help my patients and clients, this book was born. I wanted something tangible I could give people who were struggling with their sleep to put them back in the driver's seat and impart what I have learned in my more than twenty years in the field.

This book is meant to be read like a sweeping novel. It is not a reference book. I do not want you to skip ahead to the part of the book that you think is most important for you. It's *all* important! Think of it as a complete process for understanding and overhauling both your sleep and the way you think about sleep. If you do it my way, you are going to finish this book with a newfound sense of what it means to have healthy sleep.

An Introduction to
Sleep Medicine

F AMILIAL FATAL INSOMNIA IS A very rare but real condition related to mad cow disease. The afflicted individual develops a progressive difficulty sleeping accompanied by hallucinations, panic attacks, and rapid weight loss. Severe cognitive impairment begins, and eventually the individual becomes unable to speak. In the end, the individual dies because of his relentlessly progressive inability to sleep.

Relax. You don't have it.

Despite how rare this condition is, most people who struggle with their sleep feel as if it too is a hopeless situation. There are few health issues that cause more stress and anxiety than sleep problems and few that are as innocuous and treatable. As a neurologist, I have dealt with conditions that are serious and devastating. Amyotrophic lateral sclerosis, or Lou Gehrig's disease, leads to loss of muscle control, causing a slow and painful march to death. A stroke that leaves an individual unable to speak is an awful and often permanent condition that we have little ability to treat once it occurs. Sleep complications can lead to serious health conditions, but unlike many neurological disorders, sleep conditions are treatable. You can fix them.

This is certainly not meant to diminish the significance of sleep disorders. Conditions like sleep apnea, in which a patient frequently stops breathing at night, cause hypertension, diabetes, and heart failure. In 2007, sleep researcher extraordinaire Tom Roth found that insomnia may affect as many as one third of our population at any given time. Maurice Ohayon's research demonstrated that restless legs syndrome may be responsible for poor sleep quality in 5 percent or more of the adult population. Sleep disturbances can contribute to problems as varied as gastroesophageal reflux disease, mood disturbances, memory problems, and weight gain. These are serious issues and huge numbers of people are affected.

So if treatment is what you need, why are you reading this book and not gowned up on your primary care doctor's exam table getting your problem fixed? Perhaps it's because less than 10 percent of you have ever visited your primary care doctor specifically to address a sleep problem. Furthermore, according to the National Sleep Foundation, if you are not bringing it up, only 30 percent of primary care doctors ask patients about their sleep. This is shocking because we spend approximately one third of our lives sleeping. To date, I've never experienced sudden visual changes or significant rectal bleeding, but I get asked about those symptoms every time I go to the doctor. Trust me, when I suddenly see blood originating from that orifice, my doctor will know about it immediately. He won't have to ask.

Speaking of doctors, let me take you behind the scenes of a typical medical school. Regardless of a physician's eventual specialty, everyone in medical school studies everything. Medical students spend years attending one lecture after the next covering all aspects of medicine. That's why this part of medical training would not make for compelling television. In my second year of medical school, a neurologist who taught sleep medicine came into our lecture hall and told us that during the following fifty minutes we would learn about sleep disorders.

I remember the lecture well. It began with a video of an elderly couple being interviewed. The wife was in tears as the husband choked up telling the story of how he dreamed he was chasing a deer through his barn. He remembered catching the deer, and as he was getting ready to drive the buck's head into the wall of the barn, he woke up with his wife's head in his hand.

This was an example of REM behavior disorder, a condition in which the normal paralysis that accompanies dreaming is impaired. The neurologist discussed sleep apnea too, but I didn't remember that part because, like most of the other students, I was too horrified by the video I had just seen to pay attention any longer.

As quickly as the lecture began, it ended. That was the extent of our sleep training, and it may be all of the training your primary care physician has had as well. According to researcher Raymond Rosen, most physicians have received less than *two hours* of training about the entire field of sleep in their four years of medical education. Mihai Teodorescu and sleep specialist Ronald Chervin's research from 2007 revealed sleep is dramatically underrepresented in medical school textbooks. Given that our psychiatry lecture about men who fantasize about their wives' footwear lasted thirty minutes, you can see just how dramatically underrepresented the whole of sleep medicine was in our curriculum.

Despite what is often minimal education about sleep medicine, it is among the most common problems physicians are asked to address. Yet trying to treat a problem that involves anything other than an old guy assaulting wildlife in a dream might be difficult for your doctor. This is not an attack on the primary care physicians of the world. As their compensation from insurance companies declines and their malpractice premiums increase, they are seeing more patients in less time. Their patients often carry with them numerous diagnoses that require attention, making issues related to sleep an afterthought. So to criticize a primary care doctor for failing to treat sleep difficulties effectively is like being upset at a pathologist for a difficult labor and delivery—it's not her job.

So what can you do? Get smart and quit getting your sleep information from *Cosmo*, from sleep books that make a simple subject complicated, and from your next-door neighbor. It's time for you to stop complaining about your poor night's sleep and throw your misconceptions about sleep out the window. You *can* understand sleep and why yours ain't workin'. So gather up your over-the-counter sleep aids and toss them down the drain. School is about to begin.

1

WHAT IS SLEEP GOOD FOR? ABSOLUTELY EVERYTHING!

REMEMBER MAD LIBS BOOKS FROM when you were little? I used to love getting that little paper notebook with my Scholastic Reading Club book order when I was in middle school. The little tablet was filled with stories that you completed if you knew your parts of speech. A few adjectives, verbs, and names of your friends later, and you had a slightly illogical but hilarious story.

I've always thought about sleep and its relationship to other medical conditions as a Mad Libs game. When it comes to the connection between sleep and the many other things going on in our bodies, there is almost no disease or organ system in which you cannot find some kind of relationship. Don't believe me? Give the exercise a try and you'll see what I mean.

SLEEP LIBS

Fill in the Mad Libs below[1]:

Why Quality Sleep Is Important

At night, when it is _____, I like to get into my _____
                    (a time on the clock)   (adjective)

bed. It takes me no time to _____ into a _____
                                    (verb)   (adjective)

sleep. This is a good sleep because poor sleep can lead to

_____ . Scientists have shown in a recent _____
(medical condition)   (adjective)

study on human _____ that getting less than _____
                    (body part, plural)   (number)

hours of sleep at night can lead to a _____ case of _____ .
                                          (adjective)   (medical condition)

Hilarious, right? What's amazing about this Mad Libs "sleep lib" is that there are relatively few ways that you could fill it out and make the story untrue. For the "medical condition," you could have written hypertension, heart attack, stroke, obesity, diabetes, cancer, heart failure, migraine, atrial fibrillation, depression, bed-wetting, or neurodegenerative disorders and memory disturbances like Alzheimer's disease. The list goes on and on, and all of the answers make perfect sense!

As you read this book, think about sleep as one of the foundational processes within your body that you can actually change. To me, the three main pillars of good health that we can exert some control over are nutrition, exercise, and sleep. Sleep is an amaz-

---

1 Mad Libs® is a registered trademark of Penguin Random House LLC. Used with permission.

ingly important process that happens in our bodies. If you take away nothing else from this book, please understand that sleep is not the absence of wakefulness. In other words, sleep is not a light switch in your brain that is either on (you reading this book, sipping your coffee) or off (sleeping). Your body is doing amazing things at night while you sleep.

As for the workings of the brain, in addition to being a sleep specialist, I'm a neurologist, or a brain doctor, by training. Sleep specialists are often neurologists, but they can be psychiatrists, pulmonologists, internists/family medicine practitioners, and even pediatricians. Why would any lung doctor specialize in sleep? I have no idea. It seems to me that sleep has about as much to do with the lungs as it does with the kidney or spleen![2] While virtually every system and organ of the body is in some way affected by sleep, sleep resides in the brain. This is where sleep both originates and is controlled. Sleep is a neurological state, so when it comes to sleep, the brain is where it's at. For this reason, it is where we will begin surveying the impact poor sleep has on our bodies. If you think your all-nighters or your crazy shift work schedule is no big deal, you might want to sit down before you continue to read. Long-term poor sleep is like bad cosmetic surgery: risky, costly, and not pretty.

Sleep and the Brain

I remember a few things vividly from medical school. I remember the unmistakable smell of cadaver preservative and how hard it was to remove the fat from the organs we were dissecting.[3] I remember taking a test and being shown a dazzling picture of gallstones and thinking how strangely beautiful they were. I thought gallstones, polished up, would make lovely beads for a necklace.

I also remember talking about the lymphatic system, a fluid passage system in our body responsible for collecting and circulating

2 I'm still waiting for the *Time* magazine cover story "Scientists Unlock the Mysteries of the Spleen."

3 I also recall a classmate discovering that a hair dryer could be used to warm fat and make it more easily slide off the body. This created a smell that was incredibly awful. Unfortunately, our brains tie smells very strongly to memory.

waste so it can be removed. As a budding neurologist, I was really surprised when our professor proclaimed that the nervous system was devoid of such a system. *The most important system in our body has no way of flushing out waste products, yet the spleen does?* That made no sense.

Fast-forward to 2015 when it was discovered independently by researchers Antoine Louveau and Aleksanteri Aspelund that the brain does in fact have a system for removing waste: the glymphatic system. Although scientists today generally agree on its existence, it was another aspect of the glymphatic system that really grabbed headlines. Scientists discovered that the main waste product the glymphatic system is removing is amyloid beta (Aβ), the protein that accumulates in the brains of Alzheimer's patients. While that fact itself is fascinating, there's more:

The glymphatic system is 60 percent more productive when we sleep than when we are awake!

Isn't that remarkable? Not only do we have a system for pumping waste from our brain, but according to the work of researcher Maiken Nedergaard and her colleagues, the waste-removing system works far better when we are sleeping.

Knowing this, think about the long-term consequences of not sleeping well. Making the decision to stay up late at night impairs your brain's ability to get rid of toxic waste products building up during the day. Think of your brain like a massive ocean tanker. The glymphatic system is the ship's bilge pump removing the built-up water from the hull of the ship. If the bilge pump malfunctions or does not run effectively, the water accumulates and the ship sinks.[4] While this is most certainly not the full explanation for the genesis of Alzheimer's disease, it may play a significant role. A 2013 article published in the *Journal of the American Medical Association Neurology* supports this mechanism. In this study of seventy older adults, the subjects who reported either sleeping smaller amounts or having more sleep disruption were shown to have more Aβ accumulation.

---

4 Speaking of sinking ships, the investigation into the *Exxon Valdez* shipwreck/oil spill revealed sleep deprivation to be at the core of the accident. More on that later.

CUTTING-EDGE SCIENCE

MOST INDIVIDUALS THINK ABOUT GENETICS as something they have relatively little control over. If you have the genes for green eyes, there is little you can do to change that outside of colored contacts. Possessing the apolipoprotein E ε4 gene has been shown to increase an individual's risk for developing Alzheimer's disease ten- to thirtyfold over those who do not possess it. Just a few years ago, if you found out you were dealt this gene, you were pretty much out of luck. However, in a 2013 study published in the *Journal of the American Medical Association*, that idea was seriously challenged. In that study, 698 older participants were followed in a large community-based study. As part of the study, sleep quality was assessed. During the period of the study, 98 of these patients developed Alzheimer's disease. Analysis of the results indicated that better sleep quality had the ability to reduce the impact of apolipoprotein E ε4 on disease severity. The patients with a genetic predisposition for Alzheimer's disease were able to significantly delay and/or reduce their risk for developing the disease by simply sleeping better. Think about that for a minute: genetic tendencies being influenced by better sleep. We tend to think about genetic traits as being inevitable, inescapable conclusions. This study showed that our choices and behaviors can influence our bodies at the genetic level. Behold the power!

One last thing about the glymphatic system: it seems to work better when you sleep on your side. Stony Brook University researchers Hedok Lee and colleagues studying rodents found that the glymphatic system worked more efficiently when the rodent was placed on its side. One behavioral change you can implement right now that could reduce your risk of developing Alzheimer's disease is to simply sleep on your side.

Alzheimer's disease is not the only neurological disorder associated with poor sleep. A 2011 study showed a link between poor sleep and Parkinson's disease. Other neurodegenerative conditions and decreased memory function in general have been associated with poor sleep quality, according to a 2014 study.[5]

Sleep and Obesity

This is not a weight loss book. There are no fad diets or chia seed smoothie recipes at the end. Despite this, it makes a lot of sense to address the subject of sleep and obesity because up until recently, this connection was largely ignored. Looking back over decades of research, it has been apparent for a long time that increased body weight could cause poor sleep, largely related to breathing changes. This was referred to as Pickwickian syndrome, named after the Charles Dickens novel *The Posthumous Papers of the Pickwick Club*. In the book, Joe is an overweight character who tends to fall asleep frequently during the day, as many with sleep apnea do. While studies linking weight gain to poor sleep go back well over fifty years, studies linking poor sleep to weight gain are much more recent. There have been many studies over the last several years that demonstrate poor sleeping leads to weight gain. The mechanisms behind these studies vary greatly, but here are some highlights:

- Numerous studies have shown sleeping fewer than six hours and staying up past midnight to be linked to obesity. In a 2015 study looking at the habits of over 1 million Chinese subjects, public health researcher Jinwen Zhang found higher levels of obesity in people sleeping fewer than seven hours per night. Another 2015 study published by clinical psychologist Randall Jorgensen in the journal *Sleep* showed very clearly that as sleep duration dropped, waist size increased. The evidence that disturbed sleep leads to weight gain has probably reached the "overwhelmingly so" level. This study is a great one to cite when you decide to sleep in and skip meeting your friend at the gym.

---

5 Blanking on the comment I wanted to make here. Ugh, so frustrating . . . something to do with sleep and . . . you know that thing . . . it will come to me. Just keep reading.

▪ School-age children who slept inadequately (fewer than nine hours per night) and/or erratically[6] were more likely to be obese, according to a 2008 study by circadian rhythm/ endocrine system investigator Eve Van Cauter. As I watch my older kids stay up until the wee hours of the night, I'm often tempted to take studies like this into their school and ask their teachers if the ridiculous amounts of homework are worth a lifetime of fad diets and undergarments designed to conceal muffin tops.

▪ Ghrelin is a hormone produced in our gut. Ghrelin acts on our brain to promote hunger, but it also may play a key role in the pleasure associated with eating. Ghrelin makes us crave all of the processed foods on full display at twenty-four-hour convenience stores. Clinical investigator Shahrad Taheri's 2004 study showed that as sleep duration goes down, ghrelin production goes up, increasing the likelihood of overeating and obesity.

▪ Poor sleep quality can affect levels of the chemical leptin in our body. Leptin, produced by our fat cells, induces the feeling of fullness and puts the brakes on our appetite. When we sleep poorly, leptin levels are reduced, which makes us want to eat more, according to a 2015 study by researcher Fahed Hakim.

▪ Researchers Alyssa Lundahl and Timothy Nelson's 2015 study demonstrated that after a poor night of sleep, our energy levels are reduced. One compensatory mechanism is for us to eat more in an effort to boost our energy.

▪ With poor sleep come decreased impulse control and greater risk-taking behaviors. These factors could lead to eating poorly during periods of disturbed or inadequate sleep, according to Harvard researcher William Killgore in his 2006 study.

---

6 As a result of travel sports.

CUTTING-EDGE SCIENCE

A 2015 STUDY LOOKING AT 3,300 youths and adults came to a very sobering conclusion about sleep and weight. Lauren Asarnow and her group from Berkeley studied the effects of chronic sleep loss on people's weight. They showed that, over time, for every hour of sleep an individual lost, he or she gained 2.1 points on their body mass index (BMI).[7]

Sleep, Your Heart, and Blood Pressure

The effects of poor sleep are probably most damaging to our heart and circulatory system. Poor sleep quality has been shown in zillions[8] of studies to increase risk for heart attack, elevated blood pressure, heart failure, and stroke. While most of these studies center around sleep apnea, a condition that causes the upper airway to collapse and makes it impossible for the sleeper to breathe, not all studies center on this condition. Recent research has shown that any condition that fragments sleep (not just sleep apnea) has the potential to elevate blood pressure.

Atrial fibrillation is a condition in which the heart starts to beat in an unsynchronized rhythm. This is not a good thing since a coordinated heartbeat makes sure blood quickly and efficiently moves through the heart. When an individual develops atrial fibrillation, the coordinated efforts of the various chambers of the heart are lost, causing blood to pool in the heart. The swift movement of blood is one of the mechanisms that prevents clotting. When blood sits for long periods of time, clots can form.[9] When clots form, bad things like strokes and pulmonary embo-

---

7 Going to bed early: beauty sleep. Staying up late: booty sleep.

8 Give or take.

9 This is why you want to get up and downward dog in the aisle of the airplane every now and then when you fly.

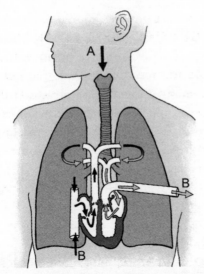

Figure 1.1. Why suffocating upsets your heart.

lisms can sometimes follow. These are not conditions you want in your life.

Guess what. Your sleep can influence whether you develop a funny heart rhythm and a massive blood clot in your leg! Studies have shown that individuals prone to atrial fibrillation can reduce the chance of the condition coming back once it's treated by also treating their sleep apnea if they have it. After individuals treated their breathing disturbance, their risk of developing atrial fibrillation dropped from 82 percent recurrence to 42 percent!

Let's think about our heart for a minute. Where does the heart live? In our chest. Who are its neighbors? The lungs. Let's look at the simple picture above.[10]

This is a picture of your heart and lungs. Notice how your heart is positioned right between your lungs and how everything is sealed up within your chest cavity. Your heart needs to be there because its main function is to pump blood that no longer has oxygen (the blue blood: blood turns blue/dark when it has no oxygen) to the lungs, where it can have oxygen put into it once again, turning it bright red. In this arrangement, the chest cavity acts like a bellows.

---

10 Wow, you know there must be a serious relationship between sleep and the heart if the author demanded the art department render a diagram of this whole affair.

For the lungs, that's a good thing. When we expand our chest, just like in a bellows we create a negative pressure, or a vacuum. There is a saying that nature abhors a vacuum, and it's true; the air outside our lungs rushes in to fill up the space created, thus making us inhale. When breathing works well, everything is fine. When an individual has trouble breathing, things get problematic. Look again at the diagram and imagine a person struggling to breathe at night. To keep from suffocating, she is going to try to suck air ("A") into the lungs with more and more force.

Unfortunately, because of the real estate the heart occupies within that chest-cavity bellows, any sucking that is produced to pull air into the lungs has the consequence of sucking blood back into the heart (the "B" on the right).

If the heart has trouble pumping blood out, the blood coming back to the heart (the "B" at the bottom of the diagram) has no place to go. It cannot move into the heart because the blood is not moving out efficiently. The blood cannot turn around and go backward. So what is the body's natural solution?

It turns out there are two consequences, and both are bad. The first consequence is that fluids are pushed out of the blood vessels and into the tissue of our body, usually our legs. This is the mechanism behind leg swelling or edema.

Second, the heart works harder to pump the blood out. What happens when a muscle, like the heart, works harder? It enlarges. This is the beginning of heart failure.

For individuals who leave sleep-related breathing disturbances untreated, the long-term consequences on the heart are devastating. Heart failure is an inevitable outcome.

Sleep and Mood

All of this talk of poor sleep, heart failure, Alzheimer's disease, and not fitting into your favorite jeans is a super downer. Want something to help brighten your mood? Try sleeping. Seriously! Poor sleep can lead to depression and negative mood consequences. This would be a great time to put on your favorite Smiths album.[11]

---

11 Heaven knows you're miserable now.

▪ Poor sleep itself can dramatically worsen mood and has been linked to worsened depression and anxiety. For some mental health professionals, the association between depression and insomnia is so strong, they don't diagnose depression in someone who does not show signs of sleep disruption.

▪ Frequent awakenings during the night, regardless of the cause, may contribute significantly to worsened mood and negative emotions. In his 2015 study, Johns Hopkins researcher Patrick Finan found the effects of interrupted sleep on mood may be more powerful than the effects of reduced sleep.

▪ Circadian rhythm disorders are frequently associated with depression and other mood disorders. As patients spend more time in bed and withdraw from typical activities, their schedule and sleep-wake cycles become a big mess. Like syndicated episodes of *Law & Order*, their eating, exercising, and sleeping happen at all hours of the day and night.

▪ For patients with obstructive sleep apnea, depression is a common concurrent condition. In a 2015 study, David R. Hillman and others from the University of Western Australia found that the treatment of sleep apnea can reduce the incidence of depression significantly, dropping it from 73 percent to 4 percent.

▪ Bipolar patients can have significant issues with sleep. Manic episodes can feature long stretches of time in which the patient is unable to sleep. A 2015 study showed depressive episodes carry with them risk of poor sleep quality, oversleeping, and sleep schedule problems.

Sleep and Cancer

I wish I were making this section up. As someone who has been in the field of sleep for as long as I have, I find the emerging association between sleep dysfunction and cancer is still very unsettling. While there is evidence that poor sleep quality may be linked to a variety of cancers (prostate, oral, nasal, and colorectal as well as primary nervous system cancer), it is the emerging link between poor sleep and breast cancer that seems to be strongest. Not only do sleep disturbances like shift work and sleep deprivation represent a potential risk factor for the development of breast cancer but epidemiologist Amanda Phipps found insufficient sleep before the diagnosis may be a predictor of treatment outcome.

In 2007, the World Health Organization (WHO) published a monograph titled "Carcinogenicity of Shift-Work, Painting, and Fire-Fighting." Let that sink in. Not only is the WHO grouping shift work with inhaling paint fumes and smoke from burning houses in terms of causing cancer, but it is listing shift work first on the marquee! In this early investigation, researchers found a relationship between shift work and breast cancer as well as a general decline in immune system functioning. Subsequent research surrounding shift work in particular has led the International Agency for Research on Cancer, an agency of the World Health Organization, to classify shift work as a probable (group 2A) carcinogen.

Sleep and Your Immune System

"Go to bed or you're going to get sick." How many times do kids get hit with this parental favorite? When I was a kid, these words had zero effect on my willingness to stay up all night watching Letterman or sorting my football cards.[12] At the time, this connection between my bedtime and my overall health made little sense to me, but in college, the combination of stress and all-nighters sometimes ravaged my body. I'm pretty sure I came down with the plague after a particularly grueling week of finals.

Why is it that a late night spent cramming or being on call at the

---

12 This was back during a time when the Raiders were awesome, and the Patriots were terrible, if you can believe such a time existed!

hospital almost invariably caused me to get sick or acquire a nasty cold? Our immune system's function is intimately tied to the amount and quality of our sleep.

■ In a 2015 study led by Aric Prather of the University of California, San Francisco, subjects were given the rhinovirus voluntarily. If the subjects receiving the rhinovirus had slept six hours or fewer, they were more likely to develop the cold versus the subjects sleeping more than seven hours.

■ Another recent study from a team of researchers from Taipei, Taiwan, showed disturbed sleep as a risk factor for developing autoimmune system disorders. These conditions can produce a wide variety of disabling symptoms, such as painful and distorted joints (rheumatoid arthritis); a slowly fusing spine (ankylosing spondylitis); dry eyes, mouth, and other body parts (Sjögren's syndrome); abnormal growth of connective tissue throughout the body (systemic sclerosis); and a condition that can cause damage to virtually any part of the body (systemic lupus erythematosus).

■ In a 2013 study, a group of college fraternity brothers was asked to share the same red plastic cup for an entire weekend of all-night underage drinking. After a raging three days and two nights of continuous partying, researchers were shocked to discover. . . .

Need I go on? I could virtually go organ to organ through the whole body and show you how lack of sleep is harmful. I mean, we haven't even gotten to the part where I tell you screwed-up sleep can devastate your body's ability to regulate blood sugar, creating a huge diabetes risk factor![13] Do I need to say anything beyond the fact that sleep problems screw up your brain and may lead to Alzheimer's? The brain is the most important organ in your body.

---

13 If you feel shortchanged about the diabetes thing, take a look at "Impact of Sleep and Circadian Disruption on Energy Balance and Diabetes: A Summary of Workshop Discussions" in the journal *Sleep*, volume 38, issue 12, 2015. After you read it, see how motivated you are to stay up late eating M&Ms.

Period. Any talk about the lesser organs is just wasting both your time and my time. Let's move on.[14]

....

CHAPTER 1 REVIEW

1. When sleep is not working properly, you don't work properly.
2. When people say scientists don't know why we sleep, they are wrong. We sleep to stay alive.

Why do we eat a crab cake? Why do we drink a glass of orange juice? Because we have no real choice. We must eat to live. With sleep, we have even less choice because when the drive to sleep is strong enough, it overpowers us and forces us to sleep. My motto is: "Sleep always wins."[15] Sleep is a powerful driver of human behavior. What else drives us? Read on and find out.

---

14 If you need more convincing, (1) wow and (2) go buy Arianna Huffington's book *The Sleep Revolution*. It will beat you over the head with how your lack of sleep is killing you a thousand different ways.

15 My motto when I started college was: "Sleep is for losers." That changed when, after weeks of staying up way too late, I inadvertently fell asleep prior to a sorority champagne slip-and-slide party and missed the whole thing.

PRIMARY DRIVES
Why We Love Bacon, Coffee, and a Weekend Nap

'VE PROBABLY NEVER MET YOU. You could be anyone skimming the pages of this book—a tired college student waiting in line at the university bookstore, a mother of three enjoying some tea at Barnes & Noble while her kids are at school, a multibillionaire syndicated talk-show host deciding this is the book her millions of viewers need to read immediately. Despite the fact that I have no idea who you are, I'm going to make several statements about you that I know for a fact are true.

1. You have eaten something in the last several days.
2. You have had something to drink in the last several days.
3. You have thought about sex in the last several ~~minutes~~ days.
4. You have slept some in the last one or two days. (If you're the tired architecture student on a three-day/two-night bender to finish your urban design and recombinant space project, you don't count.)

Before you pay for this book, consider all four of these statements. If you think any of them are untrue in your situation, put this book down and buy something else.

Food, Water, Sleep, and Sex (Not Necessarily in That Order)

The truth is, I have no special powers. My psychic secret is all about primary drives. In 1943, American psychologist Clark Hull put forward an idea called the drive reduction theory. He felt that the behavior of all organisms was governed by their goal of maintaining homeostasis, or balance, in certain primary drives. We have a need or primary drive for food and water to nourish our bodies. We have a primary drive to reproduce. And guess what. We have a primary drive to sleep. Because of this, the longer we go without sleep, the more determined our brains are to get it, ultimately to the point where it is no longer a choice. In other words, sleep is inevitable.[16]

Many of the patients I've treated come to me insisting that their problem is that they "don't sleep." They never fall asleep, or they wake up and cannot get back to sleep. Anyone who says this to me is actually suffering, at least in part, from a more fundamental problem: They are sleeping; they just don't perceive their sleep effectively. In other words, their assertion of not sleeping is just plain wrong. The medical fact is, we all sleep. It's a primary drive. The body insists on it. So the first thing I need to tell you if you are one of those people who "never sleep" is this: You need to accept one simple fact, or you will be doomed to struggle with your sleep forever.

YOU SLEEP.

Say it out loud. I don't care where you are. Are you in a library? Okay, then whisper it. "I sleep." Two words, six letters. Say it again,

---

16 You can fall asleep during your own wedding, during the birth of your own child, or during intercourse . . . all true patient stories.

"I sleep." Do you sleep well? Maybe or maybe not, but *you do sleep.* Say it, "I sleep." Do you look at the clock every hour? Maybe or maybe not, but *you do sleep.* I cannot emphasize this point enough as it is usually the first law I have to establish with my new patients. If your sleep is a geometry class with lots of axioms, postulates, properties, and proofs, think of this as Law 1: *You sleep.*

 CUTTING-EDGE SCIENCE

I CAN SENSE IT. You still don't believe that you sleep or you think you are someone who is getting only two to three hours of sleep or fewer every night. Okay, I hear you, but consider this. The really, really smart sleep researchers who wrote the text-books I used during my sleep fellowship did studies looking at the ability of humans to function getting little sleep. David Dinges and Hans Van Dongen put research subjects into groups getting four, six, or eight hours of sleep every night. They were carefully monitored to make sure that no subjects were sneak-ing off into supply closets and catching some zzz's. The study ran for only two weeks. Just two weeks! I have people telling me every day they have not slept for years. The sleep project lasted fourteen days—that's it.

In the study, the subjects had their attention measured by performing a series of psychomotor vigilance tasks. By the end of the study, one quarter of the *six*-hour group was falling asleep during the tasks! The four-hour group was doing even worse. It's interesting that, when asked, the sleep-deprived subjects didn't really think they were impaired. In other words, despite falling asleep on the computer, these individuals were emailing all of their friends, "Hey, I just did this crazy sleep-deprivation experiment. I think I crushed it!"

Dear Reader, I wanted to include a cool recent study in which sleep-deprived research subjects were given a couple of hours of sleep for weeks on end and demonstrated no discernible disability. I really did. But I reluctantly must report that this research does not exist. People simply cannot do it. They get too sleepy. They fall asleep during the studies. Basically, to sum up the current thinking among every reputable sleep scientist in the world, here is the gospel: There is probably a very small percentage of the population that can get six hours of sleep or slightly less over a relatively prolonged period of time and can maintain their performance, but performance deterioration will occur. The idea that there are people out there getting two or three hours of sleep for long periods of time and are still able to walk, chew their own food, program their DVRs, and string together coherent sentences is simply not true.[17]

If, however, you are the chosen one, I want to be the one who discovers you and receives all of the scientific accolades and awards. So, please take a moment to answer the following questions:

- Are you human? _____
- Are you free of diagnosed psychological illnesses? _____
- Have you consistently, *without exception*, slept an average of only three hours or fewer for the last year? _____
- Are you willing to buy a Fitbit and wear it around to prove that you don't sleep? _____
- Are you willing to be studied, have your picture taken, and allow Dr. Winter to parade you around to other sleep researchers so that he can achieve great fame and fortune?

If you answered yes to all of these questions, we need to talk immediately. Send your information to my publisher. We'll sort it all out from there.

I love my job. All day long I get to talk to people about their sleep. About once every couple of weeks I'm confronted with a panicked patient who is pleading for my help because he or she can't sleep at all or has gotten some ridiculously small amount of sleep in the previous week or two.

---

17 This is something I see all of the time. My favorite example was a highly successful lawyer who felt she was able to achieve only six hours of sleep *per week*! Despite her assertion, she was wide awake and felt absolutely no sleepiness.

"You have to help me. I've slept only two hours in the last fourteen days!"

What makes this type of patient so fascinating is that he will often top this statement off with the following zinger. "I wish I could nap, but I lie down during the day, and I can't fall asleep." Wow, an individual who not only cannot sleep during the night, but for some reason, despite his 336 hours of wakefulness, still feels no ability to sleep even during the day. Get Guinness on the phone! This is better than the creepy long-fingernail guy![18]

"So what do you do when you get in bed?" I ask.

"I just lie there and think about things. . . . I can't turn my mind off."

"You lie there all night doing nothing?"

"Yeah. My best opportunity to sleep usually comes between the hours of eleven and midnight. If I don't fall asleep during that hour, I miss my window, and I'm going to be up all night."

Huh?

Speaking of the *Guinness Book of World Records,* while they maintain records of virtually every kind of feat imaginable, they no longer recognize sleep-deprivation records. The current record holder, Randy Gardner, set the mark of 11 days and 24 minutes in 1964. During this trial, it became progressively more difficult for Gardner to stay awake. His brain engaged in microsleeps (brief periods of uncontrollable sleep usually lasting less than thirty seconds), and he suffered hallucinations, severe cognitive impairments, and even paranoia. This paranoia has been noted in several sleep-deprivation experiments, the most unfortunate of which was seen in the case of disc jockey Peter Tripp, whose sleep-deprivation publicity stunt of 201 hours seemed to have very lasting psychological effects (not the least of which was a belief that he was an impostor of himself).

The bottom line is that true, honest-to-God sleep deprivation is difficult. In research situations, it can be damn near impossible to keep subjects awake for even relatively short periods of time. This kind of sleep deprivation is not without short-term consequences, including an overwhelming drive to sleep. In other words, nobody

---

18 Actually it's not. Nothing beats creepy long-fingernail guy.

who attempts these stunts describes difficulty falling asleep. It is an accepted fact that true sleep deprivation (that is, you are in a situation in which no sleep is allowed to happen[19]) always leads to sleep or sleep intrusion (uncontrollable periods of sleep) and performance decline. In other words, if you are sleep deprived, you and everyone else know it! If, however, you think you are sleep deprived but you show no tendency to nod off when you are stretched out on a couch, does it really make sense to you, knowing what you know, that you really are sleep deprived?

THE DO-NOTHING-FOR-A-REALLY-LONG-TIME EXERCISE

IF YOU ARE SOMEONE WHO believes you are suffering from long-term sleep deprivation, and can't sleep no matter how hard you try, do this little exercise.

1. Eat a little something and try to use the bathroom. Sleep Exercise 1 is going to take a while.
2. Turn off your cell phone and telephone ringer and ask that your family leave you alone until the experiment is over.
3. Demand total privacy because the situation you're dealing with is "wicked serious." Say it just like that and nobody will bother you.
4. Find a comfortable, private place in your house or office to lie down.
5. Kick off your shoes, turn off the lights and lie down there.
6. *Do not sleep!* Just lie there for the next seven hours.
7. Reflect on your experience.

The Do-Nothing-for-a-Really-Long-Time Exercise is a real bitch, huh? Doing nothing for one hour is a real drag. Doing nothing for

19 In research protocols, the truly sleep deprived often practically required a cattle prod to keep them from falling asleep.

seven is terrifically difficult,[20] yet people everywhere claim to be doing it nightly when they complain that they did not sleep the night before.

To put things in perspective, not sleeping for four days and subsequently not being sleepy is analogous to an individual coming into my office saying that he has not eaten for four days and yet strangely feels no hunger *and* is gaining weight. Yes, I know that toward the end of terminal starvation, the body feels little to no pain, but you get what I mean.

Not sleeping and simultaneously not feeling sleepy go against sleep as a primary biological force. To that end, I've had patients tell me that the longer they are awake, or the longer their kids are awake, the *less* sleepy they or their children become. This makes some sense when you consider the brain process for maintaining wakefulness or vigilance as being a separate process from the one that initiates and maintains sleep. Despite this, an individual will *always* become sleepier the longer she is awake. This drive can temporarily be trumped by increased vigilance or anxiety.[21] This does not mean that the individual cannot sleep. It may mean that other factors are at play keeping her awake *in that moment*. Getting into bed and smelling smoke might increase anxiety and keep her awake. Hearing something move under the bed might keep her awake. Feeling really worried that she won't be able to fall asleep may also be a factor.

Cliff Saper, a sleep researcher from Harvard, did a study looking at the sleepiness of rats. The rats were put into cages that were either clean or dirty. Their sleep and brain biochemistry were measured. Did the rats sleep? Absolutely. Did they both sleep the same? No. The rats in the dirty cage showed more signs of hyperarousal and did not sleep as well as the rats in the clean cage. Their little rat anxiety about their dirty cage was a factor inhibiting their ability to fall asleep. They still developed sleepiness in the same way their clean partners did, but their heightened anxiety got in their way of falling asleep.

---

20 I did an amazing job of doing nothing for hours while my wife was in labor. I guess that's not entirely true, as I did cut an umbilical cord at some point, so that about evened things out.

21 Rolling the car window down, turning on the radio and singing the current Taylor Swift single can do it too.

Do you sleep well? Probably not or you would not be looking at this book. That's okay. That's different from not sleeping. The complaint "I can't sleep" is inaccurate and untrue, so stop reciting this mantra and reinforcing this in your mind-set. I have no qualms about interrupting my patients and correcting this sort of language in my clinic once I've explained this fact.

"I don't sleep. It's not just me. My mother never slept. She would . . ."

"No. Stop right there and try again."

"Uh, I have difficulty sleeping and my mother had difficulties with her sleep too."

"Much better. Continue."

Every patient who comes into my office and answers my question "What can I do for you today?" with "Help me sleep" gets a primary drive punch in the face from the get-go. Stop telling yourself (and others) that you don't, or can't, sleep. No matter how dirty your cage is, so to speak, your body will not allow you *not* to sleep. If you cannot eat or drink, you're going to die. If you truly can't sleep, you're going to die too, probably within a few weeks. I bet your sleeping problems have been going on longer than that and you're not dead. What does that tell you?

I find it interesting how patients deal with sleep difficulties compared to eating difficulties. Many individuals have arrived home at dinnertime, sat down, looked at the chicken cutlet on the table and felt like they simply were not hungry. One might pick around at it and the accompanying microgreen salad before simply deciding to skip the meal. Most people (who don't suffer from anorexia) would think little about this decision because they know in the back of their mind their appetite will return, and they will go on eating and ultimately thrive. Swap that hiccup in hunger with a temporary lull in the drive to sleep. An individual gets into bed to sleep and simply doesn't feel like it. Many people are immediately concerned about this development, and the stress it causes may prevent sleep later in the night or even later in the week. For many, the confidence that the boat will right itself when *sleep* is involved simply does not exist.

Although I often compare the brain's drive to sleep with its drive to eat, there are subtle differences. Our brains do not technically have the ability to make us eat. We can become terribly hungry and strongly driven to eat, but in extreme cases of anorexia or some kind of volitional hunger strike, an individual could overcome hunger and starve to death. With sleep, however, the brain retains the final say and actually has the power to force sleep on us all. We hope this doesn't occur when we are driving home from work.

How Much Sleep Do We Need?

Enough. That's the answer. You need enough sleep. Not too little, or else you will fall asleep at the dinner table. Not too much, or else you might find yourself in bed twiddling your thumbs, waiting to fall asleep. Neither of these activities are fun.

Reporters investigating sleep usually write one of three articles, centering around one of three questions: How can we get better sleep? Is napping a good thing? and finally, How much sleep do we need?

I am going to preface this little section of the book with some disclaimers. I don't really want to include this section, but I feel it should be addressed.[22] I also want to make it clear that these are just guidelines and not necessarily goals. As you read this section, keep in mind that sleep need is as individualized as caloric need. If you sleep well, feel well, and don't have symptoms of excessive sleepiness, whatever amount of sleep you are getting is probably okay.

Sleep needs change over a lifetime, but you probably knew that if you have spent any time around babies. Babies are little sleep-mongers and don't seem to be able to do much more than fall asleep, eat, and fuss about things—typically body waste in their diapers, hunger, teething pain, and/or gas. As time passes, and the young child starts doing more advanced things like calculus and Snapchatting, take a look at their sleep patterns. Chances are they are sleeping way less and their naps are probably nonexistent. Don't worry; this is normal.

---

22 It would be like Neil Armstrong writing a book about going to the moon and neglecting to write about how he and Buzz Aldrin fought about who kept drinking all the Tang.

According to Stanford sleep researcher Maurice Ohayon's 2004 study, all through our life, our sleep need is declining. Sometimes it declines rapidly like when an infant moves toward his toddler years and beyond. Sometimes sleep need is relatively stable. With this in mind, we can discuss sleep needs as they pertain to individual age groups. Again, these are only guidelines, so no freaking out if you are not perfect.

 CUTTING-EDGE SCIENCE

IN 2014, THE NATIONAL SLEEP Foundation sequestered a group of eighteen sleep experts. Their mission was to look at nine different age groups and make a determination about how much sleep they needed based on available evidence.[23] These recommendations, published in 2015, differed in many cases from the time ranges previously recommended by the foundation.

For the newborn group (up to three months), it was recommended they get fourteen to seventeen hours of sleep each day. Previously a range of twelve to eighteen hours was allowable. No more. Now if your baby is getting twelve to thirteen hours of sleep, you are failing as a parent.[24]

Infants aged four to eleven months have two hours taken away. Their range of twelve to fifteen hours represents an expansion based on the previous recommendation of fourteen to fifteen hours.

Toddlers (one to two years) have an hour taken away to arrive at eleven to fourteen hours.

Preschoolers (three to five years) have another hour lopped off, for a recommended ten to thirteen hours. Both toddlers and preschoolers gained an hour compared to previous recommendations.

Pretty much the same goes for school-age children (six to

---

23 Tradition dictates that the researchers seal themselves into an ancient sleep lab and deliberate incessantly until a consensus is reached. Interested onlookers know that the docs have decided on a proper sleep time for a given age group when they see the telltale white smoke rise from the chimney.

24 Just kidding. You are an awesome parent.

thirteen years). They should be sleeping nine to eleven hours before waking up, going to school, and not being left behind.

Teens (fourteen to seventeen years) lose an hour compared to their annoying little siblings and get to spend only eight to ten hours in bed.

Finally, younger adults and adults (eighteen to twenty-five and twenty-six to sixty-four years) drop an hour to seven to nine hours.

In a cruel, ironic twist of fate, older adults who have finally arrived at retirement and a childless home need only seven to eight hours of sleep, leading them to continually ask, "What do I do with the other sixteen hours of my day? There's nothing good on TV anymore."

Before we wrap up this chapter, it is probably important to address how our sleep has changed, not over a lifetime, but rather over generations. In other words, in their youth, did Grandpa and Grandma sleep a lot more than we do today? Given the amount of time they spent walking places uphill in the snow, having it much harder than kids do today, and saving up all their nickels to buy a few pieces of hard candy at the five-and-dime, it would seem they would not have had any time to sleep at all.

While it is easy to get the idea that we do not sleep nearly enough today and that we are sleeping significantly less than our ancestors, several studies would suggest otherwise. In a 2010 study, Kristen Knutson analyzed time diaries kept by research subjects from 1975 to 2006 and concluded that we are not really sleeping less intrinsically, although people do seem to be working more. This study did not seem to support the idea that members of modern society were sleeping drastically less than their counterparts from a generation ago.

The other study that seems to cast doubt as to whether we are sleeping less than we did in the past centered around hunter-gatherer cultures and their sleep patterns. In a 2015 study led by researcher Gandhi Yetish, ninety-four adults from the Tsimane (Bolivia), the Hadza (Tanzania), and the San (Namibia) were observed for a combined total of 1,165 days. The results indicated that these individuals were averaging only six hours and twenty-five minutes of sleep per night. This is on the low end of the current average in Western industrial societies. Although the report indicated that the study subjects were resting in their huts quite a bit, the reduced amount of sleep was unexpected.

So there it is. Congratulations. You are among the sleeping. I hope you are getting the right amount based on your age and cultural milieu, but don't worry too much if you are not. That's what this book is for. If you can honestly admit to yourself that realistically you do sleep some, we are in good shape. Take a minute to stretch and digest what you've just read. Like gnocchi, this chapter can be heavy for some. Take your time with this idea if you need to before blazing on into the worlds of sleepiness and fatigue.

CHAPTER 2 REVIEW

1. Animalistic primary drives include hunger, thirst, sex, and sleep. Without meeting these needs, you will die (except for sex, the lack of which results in the entire population fizzling out).
2. You sleep. You may not sleep well, but *you do sleep.*
3. The need for sleep varies from person to person and actually tends to decline as we mature.
4. Something about rats sleeping in dirty cages.

Well, it's great that rats sleep in cages, but you sometimes can't fall asleep in your comfy bed, even on nights when you are totally exhausted. How can someone be that tired yet not fall asleep? Let's talk about what it means to be sleepy. If you are too sleepy to turn the page and you're already falling asleep, then I guess we've succeeded either way!

3

SLEEPY VERSUS FATIGUED

Too Tired for Your BodyPump Class or Falling Asleep on the Mat?

'M TIRED. I'M SLEEPY. I'M whipped. I'm pooped, worn-out, blasted, wasted, drowsy, heavy-eyed, bushed, spent, exhausted, beat, zonked, *dead*. These terms are as common in my office as the patients fast asleep in my cozy waiting room.

To understand your sleep problem, you need to dissect its nature and determine if calling yourself "sleepy" is a great place to start. In this book, when I use the word *sleepy*, I'm referring to an individual who is likely to go to sleep or has a high sleep tendency.[25] This is an important definition because people often use the terms *sleepy* and *fatigued* interchangeably when in fact they do not mean the same thing. A person who describes herself as sleepy but tells me it takes her four hours to fall asleep is not particularly sleepy based on my definition. There is a low drive to sleep, not a high drive. Understanding the differences between sleepiness and fatigue will make you better educated about your own sleep issues and what you need to do to address them.

---

25 When I use the word *Sleepy*, I'm referring to the dwarf from the 1937 film *Snow White and the Seven Dwarfs*. Interesting that Sleepy is the first dwarf seen in the movie. I like to think this is a symbolic nod to sleep being the most important thing in the world.

Fatigue: "I'm Tired of Being Tired"

Imagine a football player as he walks off the field at the end of a game. He's hot, sweaty, and beaten up a bit from the ass-kicking he's just received from the other team. His head is low, and he's staggering slightly as he limps toward the sidelines. In the locker room, he may run into a reporter who wants to ask him some questions about the game and why on earth he would want to run the ball on fourth and fourteen. As the reporter grills him, the player probably won't respond, "We made a lot of mistakes, for sure. As the fourth quarter started, some of the guys and I started getting really sleepy. In fact, on that interception in the third quarter, I ran the wrong play because I actually nodded off for a few seconds in the huddle and never got the play the coach wanted to run. I must have dozed off several other times, too, because I don't remember a number of other plays. [*yawn*] Excuse me—I'm going to take a nap before the press conference."

Most individuals in this situation would not be sleepy, as in this example. They would be *fatigued*—they'd describe their body's energy level as low. You might go to bed when you're fatigued, but not necessarily sleepy. You climb into bed early, feeling all your strength drained away. Despite your fatigue, though, you'll actually struggle to fall asleep because you aren't sleepy. This is a recipe for insomnia.[26]

---

26 For an extra kick, add two cups of coffee, a half bar of dark chocolate, and an enuretic puppy. Mix well and serve warm while watching *The Shining.*

MEASURE YOUR OWN FATIGUE EXERCISE

THE FATIGUE SEVERITY SCALE (FSS) is a validated assessment of an individual's fatigue. Respond to the following statements about fatigue, to see what role low energy is playing in your life.[27]

During the past week, I have found that:

Strongly Disagree ←→ Strongly Agree

My motivation is lower when I am fatigued.	1 2 3 4 5 6 ⑦
Exercise brings on my fatigue.	1 2 ③ 4 5 6 7
I am easily fatigued.	1 ② 3 4 5 6 7
Fatigue interferes with my physical functioning.	1 2 3 4 5 6 ⑦
Fatigue causes frequent problems for me.	1 2 3 4 5 ⑥ 7
My fatigue prevents sustained physical functioning.	1 2 3 ④ 5 6 7
Fatigue interferes with carrying out certain duties and responsibilities.	1 2 3 ④ 5 6 7
Fatigue is among my three most disabling symptoms.	1 2 3 4 ⑤ 6 7
Fatigue interferes with my work, family, or social life.	1 2 3 4 5 ⑥ 7

Average your score. If your average score is 4 or higher, your battery is not recharging. Get some rest (not necessarily some sleep, but some rest)!

Cell phones have a wonderful little battery icon on them to let us know how much charge they have. Mine even turns red and displays an exclamation mark when it is getting critically low. People unfortunately do not have such clear indicators of low energy,

---

27 If you are simply too fatigued to do this exercise, go ahead and give yourself a score of 7 and take a little rest.

so we have to look for other clues when it is time for us to recharge. Are you having trouble finding the motivation to go to your spin class? Are you struggling to finish up some reports at work? Are you lacking the drive to simply pull a load of laundry out of the dryer and fold the clothes? These may be your body's red light indicating you are feeling fatigued.

I tell my patients all the time, when you are fatigued, get some rest. When you are nodding off, get some sleep. So what should you do if you are nodding off reading this section? Get some sleep, come back later, and continue reading.

A final thought about fatigue. It is very easy to struggle through your day with fatigue and point an accusatory finger at your sleep, saying, "If I could simply sleep more or sleep better, I would feel better during the day." Maybe.

The list of things that cause fatigue is endless. Here's a small list I've made for starters:

Hypothyroidism	Parkinson's disease
Vitamin B_{12} deficiency	Medication side effects
Iron deficiency	Malnourishment
Anemia	Pregnancy
Low testosterone	Urinary tract infection
Depression	Diabetes
Multiple sclerosis	Heart disease
Lyme disease	Glandular fever

I could go on indefinitely. There are many items I could have added to the list, including conditions like chronic fatigue syndrome. The point is this: A sleep disturbance may be the reason you wake up feeling like old boot leather every morning or it may be related to something outside of your sleep altogether. Don't get too hung up on the idea that "If I could only sleep better, then I would feel better." You may feel tired from a cause other than lack of sleep or poor-quality sleep. Understanding your sleep and solving any issues you have are the first steps to figuring out the cause of your fatigue. If you can improve your sleep with this book, and you're still tired all the time, that's something your doctor will want to know.

How do you figure out if you are sleepy? Funny you should ask . . .

Sleepiness: "I'm Not Sleeping; I'm Just Resting My Eyes" and Other Falsehoods

Sleepiness is a huge problem in this country. Small examples are everywhere: churchgoers falling asleep during a Sunday service, a doorman nodding off in the lobby of a hotel, and a geology student asleep during a riveting slide show of igneous rock. While these examples are relatively mundane, consider the following example:

> About 00:09, on March 24, 1989, the U.S. tankship *Exxon Valdez*, loaded with about 1,263,000 barrels of crude oil, grounded on Bligh Reef in Prince William Sound, near Valdez, Alaska. At the time of the grounding, the vessel was under the navigational control of the third mate. There were no injuries, but about 258,000 barrels of cargo were spilled when eight cargo tanks ruptured, resulting in catastrophic damage to the environment.
>
> The Safety Board concludes that the third mate could have had as little as 4 hours' sleep before beginning the workday on March 23 and only a 1- to 2-hour nap in the afternoon. Thus, at the time of the grounding, he could have had as little as 5 or 6 hours of sleep in the previous 24 hours. Regardless, he had had a physically demanding and stressful day, and he was working beyond his normal watch period.

This is an excerpt from the March 4, 1990, report by the National Transportation Safety Board marine accident report of the grounding of the U.S. tankship *Exxon Valdez* (the cause of the nation's biggest ecological disaster since Three Mile Island).[28] In this report, sleep deprivation and fatigue were listed as causative elements of the crash.

The story of the *Exxon Valdez* is not an isolated incident. Accidents are happening all the time both on a large scale (the *Chal-*

---

28 Three Mile Island's report also lists sleep deprivation as a probable cause.

lenger disaster) and a small scale (falling asleep bowling with your coworkers and waking up with a creative Sharpie mustache).

What makes you sleepy? For me it's the musical *Cats*.[29] I sat down to watch the performance and fell asleep so fast that I thought my wife had drugged me. In the real world, there are three clear causes of sleepiness. Certain drugs can make you sleepy. Apart from drugs, sleep deprivation and sleep dysfunction are the most common causes of sleepiness. Here's a step-by-step guide to becoming sleep deprived:

1. Purchase the first season of *Breaking Bad*.
2. Stay up too late watching Walt White slowly change from mild-mannered high-school chemistry teacher to ruthless drug kingpin Heisenberg.
3. Panic when the clock reads three hours before you need to be up for work on Monday.
4. Go to sleep.
5. Feel like crap on Monday as you limp along beating yourself up for the TV bingeing.
6. Vow to go to bed early tonight.
7. Repeat steps 1 to 6 until *Breaking Bad* is over and then replace with *Mad Men*.

It really is just that easy. There are other methods—like holding down more than one job, being at boot camp, getting up at night to feed your newborn, pulling an all-nighter to cram for college finals, getting through a neurology residency, or worrying about how everything in your busy life is going to get done the next day. The possibilities are endless, and the results are all the same. You are not averaging enough sleep for your brain to function correctly during the day; and so, like a drunk teenager, your brain essentially takes the car keys from your free will, and says, "You're pathetic. I'm driving now and you're no longer in control of when

---

29 In fairness to the wonderfully talented cast and crew of *Cats*, I was a sleep-deprived third-year medical student at the time I saw the show at the fabulous Fox Theatre in Atlanta, Georgia. Having stayed up the night before during my OB rotation probably led to my swoon during Rum Tum Tugger's introduction. Maybe if I were more of a cat guy I could have fought my way through it, but the seat was so comfortable and my wife's shoulder made such a nice pillow that I couldn't resist.

we sleep." With that, your brain starts to act like a real jerk, demanding sleep all the time. Now you're falling asleep in waiting rooms, while you're driving, during intercourse, and in all kinds of interesting situations.

Why are you falling asleep all the time? Because sleep is a primary drive, and your brain will go to great lengths to make sure it's satisfied. The lack of sleep creates a higher drive to sleep (sleepiness) just like the lack of food creates a higher drive to eat (hunger). So it follows that if the lack of sleep can make you sleepy, sleep that's dysfunctional can too. To continue the food analogy, if lack of sleep is starvation, dysfunctional sleep would be trying to live off of Milwaukee's Best, Slim Jims, and Tic Tacs.

The Holes in Your Sleep

Imagine if you developed a quarter-size hole right behind your chin that connected with the inside of your mouth. Then imagine that you were sitting in a restaurant eating. As you chewed your food, instead of it slipping neatly down your throat to your stomach, where it could be digested and eventually absorbed, the food would fall to the floor. Perhaps customers across the room don't notice the hole and only see you continuously eating, hour after hour. Eventually, an observant customer might work up the courage to ask why you never stop eating, and you'd say, "Because I'm starving!" You are in a position to eat, but the act is not complete. That's how screwed-up sleep works. It may look from across the room like the individual is sleeping okay, but get a little closer, and you'll see the holes.

Sleep apnea, for example, is a condition in which the sleeper awakens over and over because of breathing-related difficulties. He may wake up for an incredibly brief moment, so brief that his brain has no awareness of the awakening. But these awakenings can fragment sleep so terribly that sleep's restorative effects are all but negated—leaving the sleeper as sleepy in the morning as he was when he went to bed.

Thus the downward spiral begins: fragmented sleep leaves you sleepy and needing sleep (primary drive, remember)—a demand you can't satisfy because your sleep quality is so poor.

Initially, for mild sleep dysfunction, you can compensate for the reduced quality of sleep with increased quantity. Getting a few extra hours of sack time or stealing a nap during lunch may be enough to satisfy that pesky sleep drive and allow you to make it successfully through your day. But if that extra sleep becomes dysfunctional, more sleep won't make a difference. Finally, your sleep is so poor that you feel like you could sleep for a week and not feel rested.[30] You can put as much water into your gas tank as you'd like, but it doesn't mean your car will run.

A word of comfort: Scientific studies have established that if an adult sleeps well, usually six to seven hours will suffice. Many of my patients with sleep problems often feel they need nine hours or more of sleep to feel okay. Imagine asking one hundred adults how much sleep they need to feel their best. Most of the adults who unknowingly have sleep problems answer with unusually high numbers. This is one of the factors that can lead survey analysts and researchers to conclude that the average amount of sleep we all need is eight hours. Six to seven hours is absolutely fine for some people. For adults over the age of sixty-five, as little as five hours could be appropriate according to the 2015 consensus paper by Nathaniel Watson and his colleagues. The hunter-gatherer cultures mentioned in the last chapter were averaging only about six and a half hours of sleep each night and yet they appeared to be quite healthy and well adjusted.

In our culture there are probably about 40 million people who are chronically ill with sleep disorders. Bottom line: There is a tremendous amount of dysfunctional sleep out there, causing excessive sleepiness, and people cannot always "sleep more" to fix the problem.

That is the difference between sleepiness and fatigue. Now is a reasonable time to figure out if you are in fact sleepy or not. This is an easy question to ask but might be a bit tricky to answer. Understanding and quantifying your level of sleepiness are essential components to enhancing your sleep and the way you approach its improvement.

----

30 There are countless patients in my clinic who admit to using their entire weekend to sleep in an effort to accumulate enough energy to make it through the upcoming workweek. In some cases, these individuals will use vacation time for extra sleep time. With all of the time spent in bed, you can imagine what the state of their homes is like. This is a dinner party invite to avoid.

THE SLEEPINESS
SELF-ASSESSMENT

☐ Are you currently sleeping? Give yourself 3 points if you
are![31]

☐ Are you struggling to stay awake as you read this book?
Included in this would be rereading the same paragraph
over and over or reading two to three pages and
realizing you have no idea what you just read. Give
yourself 1 point if you are!

☐ Are you struggling to stay awake as you read this book?
Included in this would be rereading the same paragraph
over and over or reading two to three pages and
realizing you have no idea what you just read. Give
yourself 1 point if you are!

☐ Bad joke.

☐ Are you tuning into your favorite *CSI* show and missing
the thrilling conclusion because you are falling asleep
before the team can determine the culprit? Give yourself
1 point if you are!

☐ Are you falling asleep during sex? Give yourself 1 point if
you are and 2 points if you have a partner!

☐ Are you falling asleep in public? Give yourself 1 point if
you are!

☐ Are you falling asleep while you eat? If so, you get a
point, but forget the points, videotape yourself, and send
a copy to *America's Funniest Home Videos*; they *love*
stuff like this, and you might just win some money!

☐ Are you falling asleep during conversations? If this is with
your spouse, no points. If it's with others, collect 5 points.

☐ Are you struggling to stay awake in a car? Give yourself
1 point if you are! Are you the driver? You win! Collect 20
points and proceed directly to Boardwalk, which you
may purchase if it's not already owned.

---

31 My book worked! It put you to sleep! You are so welcome! Careful mixing it with
alcohol in the future, as it appears you are quite sensitive to this book's effects.

Sleep doctors often ask patients questions like those in the "Sleepiness Self-Assessment" to get a sense of how sleepy they are. Patients often lie during this questioning. It's okay. I'm used to waking people up from my waiting room and seven minutes later asking them, "Ever fall asleep in public?" With a straight face, they respond no. This is why I strongly suggest spouses come along to inject some reality into these appointments.

The Epworth Sleepiness Scale is an eight-question survey that attempts to objectively assess an individual's sleepiness and rate it on a scale of 0 to 24. The more likely he or she is to fall asleep, the higher the point total; most doctors view a score of 9 or 10 or higher as being excessively sleepy.

EPWORTH SLEEPINESS SCALE

Chance of falling asleep:

Situation   **Points:**

none (0) mild (1) moderate (2) severe (3)

Lying down to rest when circumstances permit

Talking to someone

Reading a book, magazine, newspaper

Watching TV

Sitting quietly in a public place[32]

Sitting quietly after lunch without alcohol

Passenger in a car for an hour without a break

Driver of a car parked at a stoplight or in traffic

                                                        TOTAL

By determining how sleepy you are, you can start to gain insight

32 Remember those patients fast asleep in my waiting room? I love it when they answer 0 for this question minutes after I have to wake them up to bring them back to my exam room!

as to whether your sleep problems might be related to your sleep quality or sleep quantity. If you are scoring high on assessments of sleepiness, that needs to be addressed. However, if you are not sleepy, it doesn't mean you don't have a sleep problem; it simply means we need to concentrate our efforts in other directions, including issues surrounding sleep quality—sleep scheduling, sleep hygiene, and sleep perception as well as the structure of your sleep. Other outside influences such as mood disturbances (anxiety, depression), diet, medications, and other medical conditions must also be considered.

 ## CUTTING-EDGE SCIENCE

Imagine you and your partner are at a party and a flirty acquaintance strikes up a conversation. As you politely converse, your partner gives you "the look," which basically means "wrap it up or prepare for a painful car ride home." A 2015 study published in the *Journal of Neuroscience* found that sleep deprivation may impair your ability to properly read facial expressions. In other words, not getting enough sleep might make you misinterpret threatening looks and lead to a difficult night sleeping on the living room couch.[33]

The couch scenario gets worse. The lack of sleep coupled with the frustration of sleeping on the uncomfortable living room couch can lead to a loss of emotional control. In another 2015 study, Talma Hendler noted that sleep loss was associated with a lowered threshold for "emotional activation." In other words, instead of simply apologizing and taking your lumps on the old sofa bed, the sleep loss creates a shortened fuse in your brain, which now thinks it's a good idea to loudly engage with your partner over why you should be in the bed and she should be in Sofa City. The ending to this conflict is too graphic for this text. Sleep tonight. Save your relationship.

---

33 You have been warned.

Why and *How* We Sleep:
The Homeostatic and Circadian Systems

Now that you're an expert on sleepiness, what causes it, and how it affects people, it makes sense to understand how your body creates sleepiness and the chemical factors that influence it.

There are two main systems in your body that work to produce sleepiness: the homeostatic system and the circadian system. These systems ideally work in concert to produce sleepiness in a way that promotes healthy and fulfilling sleep.

Homeostasis refers to bringing balance or equilibrium to a system; it's in charge of bringing rest to a system that is not at rest. The longer you go without sleep, the more powerful the drive to sleep and get your system back in balance. In the same way, the longer you read this compelling chapter and ignore the urge you have to urinate, the stronger the drive to pee will be until it's overwhelming and you can't concentrate on anything you're reading— again, the drive to balance.

A chemical called adenosine mediates the homeostatic system of sleep.

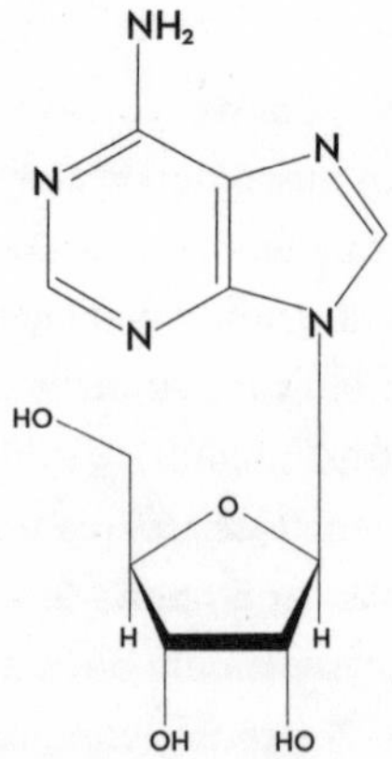

Figure 3.1: Obligatory picture of adenosine chemical structure.

As you're awake for longer and longer periods of time, more adenosine collects in your brain. Because adenosine induces sleepiness, the longer you're awake, the more likely you are to be sleepy. This is the chemistry behind sleep being a primary drive.[34]

Caffeine blocks adenosine. Ever wonder why Red Bull makes you feel so awake? Caffeine, baby, and lots of it (about 80 milligrams per can or 9.64 milligrams per fluid ounce). Need more? Try a Starbucks double shot at 20 milligrams per ounce (mg/oz) or an espresso at 50 mg/oz. Some of the newer extreme energy drinks can reach levels of more than 100 mg/oz.

Why are these drinks so totally awesome and effective? When you're awake at 4:30 A.M. organizing your wrapping paper or balancing your checkbook without some Rabid Opossum Dangerous Energy drink,[35] you'll understand how stupid that question is. These drinks temporarily block the effect of all that accumulating adenosine on that poor noodle of yours that's screaming for you to put down the remote and go to bed. New research reveals that in addition, that cup of joe may also disrupt your brain's timekeeping ability. The caffeine can act to convince the brain that it is not as late as it really is, making an individual less sleepy when they go to bed. More on sleep timing coming up.

Physical activity also increases adenosine, so the harder you exert yourself, the more likely you are to be sleepy. Exercise is a vital part of any sleep program, with hard work often being a fantastic tool to combat occasional difficulty sleeping. We'll talk more about that in Chapter 6.

Adenosine and homeostatic drives are only part of the picture. It's no accident that most of us love to sleep at night and prefer to be awake during the day. Light (most often from the sun) plays a huge role in our sleep. Ever wonder why? Think we're genetically and evolutionarily designed to seek out a killer tan? Not really.

Think about the accumulation of adenosine in the brain. If adenosine were allowed to accumulate in the brain unchecked, we would be pretty sleepy by lunchtime and really a mess by 4:00

---

34 A long time ago, they did an experiment by taking spinal fluid full of adenosine out of a sleepy dog and putting it into a well-rested dog. The infusion made the wakeful dog feel sleepy.

35 And like an opossum, you will be up all night.

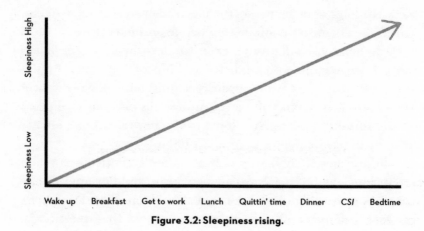

Figure 3.2: Sleepiness rising.

P.M. This drive to sleep is sometimes referred to as homeostatic pressure.

This is not how sleep works. In reality, for the most part our sleepiness level is not appreciably different at 9:00 A.M. from that at 9:00 P.M. How does that work? What other factors are involved in keeping our sleepiness levels low throughout the day?

Species survival depends on many things, the least of which is finding food. Imagine you are a flower. Which flower would you be? I think you would be a poppy. There you are, flouncing around in a field with your fellow poppies. As the sun rises, your petals open, soaking up the sun and a bee's occasional pollination. As the sun goes down, you close up protectively for the night. Day after day, year after year, century after century, little changes. These adaptations that living organisms have made in relation to the passing of time are not only crucial for survival; they are highly conserved from generation to generation. Ever wonder how important they are? Take a poppy and put it in a greenhouse closed off from outdoor light. Establish a twelve-hour on and twelve-hour off light cycle. The flower will thrive. Suddenly turn the lights on and off at random times, keeping everything else the same. Even if the amount of light remains the same, the random timing of the lights will significantly disrupt the natural rhythms of the flower, and it will die. The linking of sunlight and its day-night cycle with a biological rhythm is the basis of circadian rhythms.

In humans, this rhythm is facilitated by a chemical that is dif-

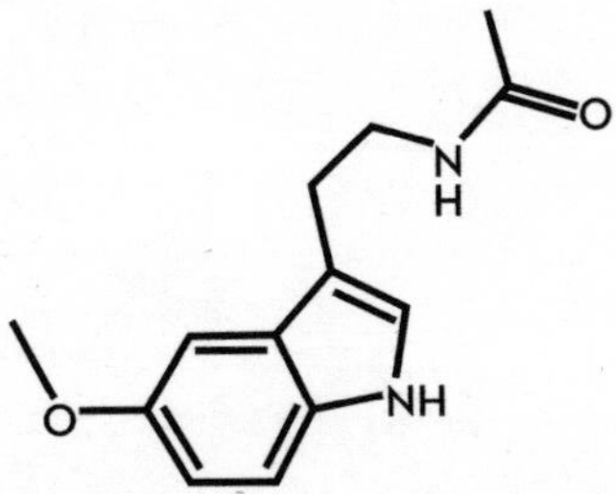

Figure 3.3. Melatonin. Notice its bitchin' benzene ring.

ferent from adenosine. It is called melatonin, and I'm certain that many of you reading this book are taking or have taken it to help you sleep at some point.

Melatonin is produced in conditions of darkness. When your eyes (retina) see darkness, a collection of cells called the intrinsically photosensitive retinal ganglion cells (ipRGCs)[36] is responsible for receiving the signal and sending it to the suprachiasmatic nucleus (SCN), the brain's timekeeper. It's the suprachiasmatic nucleus that prompts the pineal gland, a little pea-size gland in the brain, to release melatonin. Because melatonin makes us sleepy, we tend to feel sleepier at night and more awake during the day. It is interesting that raccoons have the opposite response to melatonin, which is helpful because their survival depends on sneaking around at night to find food in trash cans.

Located within the SCN, the circadian pacemakers of the brain work to counteract the buildup of homeostatic sleep pressure occurring during the day. This system modifies the homeostatic pressure curve to make it look like that shown in figure 3.4 on page 50.

Now, the relentless homeostatic drive to sleep is kept in check later in the day so you can get some things done. However, as bedtime approaches, the SCN can no longer keep a lid on things, and the big release of sleep-inducing melatonin occurs. Sleep soon follows. Notice, however, the little peak in sleepiness after lunch that happens immediately before the circadian rescue. That spike in sleepiness is the reason it is so temping to sneak a little nap after lunch. In fact, in some cultures, a postlunch siesta is the norm, not

---

36 Don't blame me. I didn't name the cells. I would have called them sleepy cells.

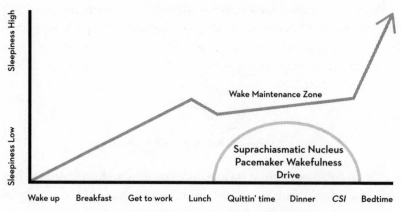

Figure 3.4: Circadian pacemakers save the evening!

the exception. Are these cultures right to give in to that urge to nap every day? Some scientists think so. I think a nap in the afternoon is fine as long as it does not affect your ability to sleep at night. You also might want to check with your boss to make sure the nap is okay.

Two systems: homeostatic and circadian. These are the underpinnings of our sleep. These chemical reactions account for the behaviors we associate with sleep and sleepiness. They are evolutionarily very sophisticated systems and are highly preserved across humans and animals alike.

CUTTING-EDGE SCIENCE

THE HUMAN "SWITCH" THAT TURNS on sleep may have been discovered in a fruit fly. Work done by Ravi Allada at Northwestern University in 2001 may be at the center of what turns sleep on and off in the suprachiasmatic nucleus of the brain. He found that when a group of neurons exhibited high sodium channel activity, the cells were turned on, causing wakefulness. When these same neurons exhibited high potassium activity, the cells were turned off, allowing sleep. This "bicycle pedal" mechanism could hold future promise for helping unlock greater understanding of sleep.

With these fabulous systems in place and presumably working well, what could possibly be wrong with your sleep? Most likely, these systems are working just fine, but you are disrupting them in some way. Let's learn more about what sleep is and you can figure out how to sleep better.

CHAPTER 3 REVIEW

1. *Fatigue* refers to a lack of energy, rather than a desire to sleep.
2. You can be fatigued or sleepy or both. You could be neither. If you are neither, why are you reading this book? What you really need is a book to help you understand why everyone hates you and your fresh and energetic life.
3. True sleepiness is caused by one of three things: medication, sleep deprivation, or sleep dysfunction.
4. Our sleep is based on two systems, the homeostatic and circadian.
5. You are or you are not sleepy. If you were brave enough to take some sleepiness assessments, you should have a feel for how sleepy you are or are not.

Congratulations. You are making progress. I hope you are learning about your own sleep and simultaneously cutting loose all of the bad information you've accumulated over the years about your sleep. Feel yourself letting go of the notion that there is something terribly wrong with your brain that's preventing you from getting a good night's sleep.

You are a good person.

And you absolutely *can* sleep.

I'll help.

SLEEP STAGES
How Deep Can You Go?

TAKE A DEEP BREATH BEFORE you launch into this chapter. People have so many weird notions about stages of sleep that this section can be mind-blowing. People use terms like *deep sleep* and *REM sleep* all the time without really having a clue what in the hell they are talking about. This chapter is intended to educate you about sleep so you never walk into your doctor's office and make the statement "I'm having trouble with migraines lately and I think it's because I'm not dreaming enough in my deep sleep. Can you help me?" In fact, by the time you finish this chapter, you'll understand why that's a truly silly thing to say.

Why is it that when you mention your night(s) of no sleep to your spouse over breakfast, you get a funny look? I see this all the time when a patient comes with his or her spouse. (Actually I often ask significant others to come into the exam room with the patient to provide another perspective.)

The patient will say to me, "I haven't slept at all for the last four days."

If I watch the face of the bed partner, I often see a funny expression, usually a subtle but confused grin. When I see that look, I'll often say something like "Why did you make that face?"

Said partner will reply (if bold enough and paying attention) with something like "Looked like you were sleeping to me when I came to bed" or "Was the grating snoring sound you were making just an act to make me think you were asleep?"

What follows is usually an awkward silence and looks of confusion on both of their faces. A polite argument can follow in which the patient describes events that occurred during the night complete with exact clock times as a sort of proof that sleeplessness did in fact occur.

This kind of speech is often countered by "Well, you were snoring like crazy both times I got up to pee last night and you slept the whole time I was getting dressed for work this morning. That's all I know." Arms cross.

Sleep perception and sleep reality are not always related. To some degree, stages of sleep and an individual's perception of sleep often constitute whether an individual considers himself a good sleeper, a light sleeper, or in many cases a poor sleeper. These perceptions are often strongly tied to the stages of sleep. These stages of sleep and the way an individual moves through them are important, so let's learn a bit about them.

CUTTING-EDGE SCIENCE

THE LABELS PATIENTS USE TO describe themselves as being good sleepers or bad sleepers are not trivial. In fact, Iris Alapin and others have demonstrated that how we view ourselves as sleepers and the labels we give ourselves may be more predictive of daytime dysfunction than our actual sleep quality. In other words, if you are a poor sleeper who views yourself as being a confident, good sleeper, you may function as well during the day as a person with far better sleep quality.

I'm amazed by how much false information is accepted as fact by patients who seek help for sleep problems. Terms like *deep sleep* and *dream sleep* are often used with little understanding about what they actually mean and what their physiological function is. In fact, most people think they are the same thing.

To clarify why equating deep sleep with dream sleep is a misconception, take a look at some illustrations.

Living people are either awake or asleep.[37]

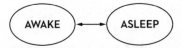

Figure 4.1. Pick a state of being, any state of being.

Sleep itself has three important phases. The foundation state is *light sleep*. Notice how light sleep serves as the passageway between wakefulness and deep sleep. *Deep sleep* is our most restorative sleep but you have to pass through light sleep to get there.

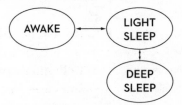

Figure 4.2. Sleep divided.

The third kind of sleep is *dream sleep* (REM sleep). Dream sleep is when the vast majority of dreams occur. (Some dreaming might occur at other times, which is discussed later in this chapter.) A few classic examples of REM dreams are listed in figure 4.3 on page 56.

Notice that light sleep is the doorway to both deep sleep and dream sleep. It is not common to go from wakefulness directly to dream sleep or from wakefulness directly to deep sleep. Furthermore, it is uncommon to transition directly from deep sleep to dream sleep. Doing so can be a clue to determining what is wrong with someone's sleep. We'll talk about them later in the book. For now, I'm trying to keep it simple, and I just want you to understand that sleep has three distinct phases. If you get that, you're way ahead of most people.

Individuals cycle through the stages of sleep in a very predictable fashion during normal sleep. We sometimes represent the transitions among stages during the night by a graph called a hyp-

---

37 A third category of "zombie" could be included as they are neither technically fully alive nor fully dead. While zombies are beyond the scope of this book, it is worth noting that there is much debate within the zombiephile community as to whether or not zombies sleep. The consensus is that they most likely do not, but probably exist in a low-energy quiescent state resembling sleep when they are not chasing humans.

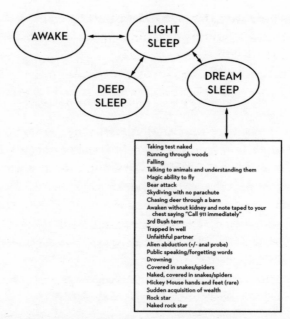

Figure 4.3. The big picture (with a bonus list of the author's favorite dreams!).

nogram. A hypnogram (figure 4.4) often represents an ideal night of sleep (in other words, it never happens this perfectly).

Let's work our way through the graph. In this example, the individual is awake for a while before drifting off into light sleep for a brief time. While the line passes dream sleep en route to light sleep, the individual is not in dream sleep because the line does not move horizontally.

This graph is the convention scientists use to trace the path of sleep through the night. In this way, it is easy to see how sleep is based in light sleep, and typically either moves to dream sleep (up)

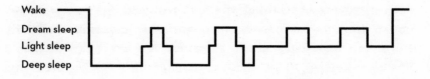

Figure 4.4. A simple hypnogram tracing an individual's cycles of sleep during one night.

or deep sleep (down). The continuity of this cycling is extremely important for sleep to work and for you to stay awake during your next staff meeting. Each of these stages of sleep has specific functions and thus specific consequences if they are disrupted.

Dream Sleep

In the early 1950s, Eugene Aserinsky, a graduate student at the University of Chicago, noticed peculiar eye movements in the sleeping children he was observing. He shared this observation with his adviser Nathaniel Kleitman, who confirmed their presence with observations of his own daughter. Unlike James Watson and Francis Crick, who took a lot of credit for discovering DNA from their colleague Rosalind Franklin, Kleitman was a class act and didn't screw Aserisnky. He gave his student full credit for being the first to notice the eye movements that characterize REM. These movements and the stage of sleep they marked would be dubbed rapid eye movement sleep, or REM sleep.[38]

Using electrodes to measure brain activity, eye movements, and muscle activity, Kleitman and Aserinsky investigated REM sleep using techniques that would later form the basis of polysomnography, the modern sleep study. Using these devices, the researchers demonstrated that the brain activity during REM sleep mirrored that of wakefulness.

Further studies demonstrated that muscle activity during REM sleep was minimal, which clearly differentiated this state from wakefulness when muscle activity is at its peak.

Aserinsky and Kleitman's later studies involved waking subjects during REM sleep. Approximately 70 percent of the subjects who woke from this sleep period reported dreaming. While new theories of sleep suggest that dreaming can occur in deep sleep, REM sleep is, for all intents and purposes, dream sleep.

Individuals usually spend about one quarter of their night in

---

38 People often ask me about the relationship between REM (dream sleep) and R.E.M., the quintessential alternative rock band. Michael Stipe, their lead singer, picked the name out of a dictionary and decided to add periods after each letter. When writing about the sleep stage, REM is written without periods.

REM sleep. REM sleep comes in twenty- to forty-minute cycles, usually about four to five times a night. The cycles typically become longer in duration as the night progresses so the longer cycles occur during the second half of the night. The longest cycle usually terminates around the time an individual awakens in the morning. This is why dreaming is most common right before you drag yourself out of bed; that's the time of your night's longest dream cycle.

DREAM SLEEP EXPLORATION EXERCISE

IF YOU ARE AN INDIVIDUAL who keeps a fairly consistent sleep schedule (that is, you go to bed at or about the same time every night and wake up about the same time every morning [weekends too, ideally]), this exercise is for you. If you aren't, enjoy your freewheelin' schedule while it lasts. My sleep schedule chapter is but pages away.

For this exercise, you will need a piece of paper and a pencil. If you are a bit more high-tech than that, have your Facebook page ready to go. You are going to have a lot to add to your wall.

1. Set your alarm thirty to forty-five minutes earlier than your typical wake-up time.
2. Go to sleep.
3. When your alarm wakes you up, were you in the middle of a dream? Chances are that if your sleep follows a typical progression, that early awakening will occur in your biggest REM cycle of the night. Furthermore, when individuals are awakened out of dreaming, they typically remember their dreams. This illustrates an important point: When an individual states that he doesn't dream, one of two things is happening: He truly does not dream, or he dreams, but does not remember dreaming.
4. I hope that after a day or two of this exercise, you'll awaken during some strange dream in which a friend

you haven't thought about for ten years is helping
you fix a flat tire. That's where your Facebook account
comes in. Look the person up . . . give him a poke.
Share with him your bizarre dream and the pictures of
you and your kids visiting the Wizarding World of
Harry Potter last summer.

Extra credit: Keep your alarm set at this early time for the next
few weeks. Do you notice how over time you awaken during a
dream less and less? This is your brain adjusting to your little
game and advancing your REM cycles to compensate. Your
brain does not like to awaken during REM sleep, so it has taken
measures to prevent it. Once this happens you can give yourself
an A++ for completing this exercise and set your alarm back to
the original time. You need your sleep . . . and how many times
can you dream about meeting Leonardo DiCaprio in the produce section of Whole Foods?

Dream sleep typically begins approximately ninety minutes after an individual falls asleep, commonly after a brief stay in light
sleep and a cycle of deep sleep. The time between sleep onset and
REM onset is termed REM latency. Measuring REM latency during
a sleep study can be useful. A shortened REM latency can be seen
in patients who are sleep deprived, suffer from clinical depression,
or who have narcolepsy—a rare condition that causes excessive
daytime sleepiness and, in some, dramatic episodes of weakness
called cataplexy. A prolonged latency is often seen in individuals
who consume alcohol or other REM-suppressing medications.

The purpose of REM sleep is poorly understood. For years,
REM sleep has been believed to be essential for memory process-
ing. This may explain why researcher Andrew Tilley found in 1978
that dreams can be difficult to remember if they are not written
down. Later, researchers would demonstrate that disruptions in
REM sleep can lead to other cognitive difficulties besides memory
difficulties, including attention problems, poor concentration,
and potential mood disturbances. Sleepiness is not classically associated with REM disruptions.

One of the most unusual functions of REM sleep may be in the regulation of pain perception. In the past, most individuals have associated pain with poor sleep.

Pain → Poor Sleep

The relationship, when stated in this direction, should not be terribly surprising. When one is in pain, one sleeps poorly.[39]

Studies have examined the reverse relationship in an effort to see if poor sleep could in fact lead to pain.

Poor Sleep → Pain

In these studies, patients were allowed to sleep in a variety of situations. In multiple studies, sleep conditions that involved REM sleep deprivation were shown to increase the levels of pain experienced by volunteers who were healthy and pain free before the studies. Study participants were monitored to determine what stage of sleep they were in at any given moment during the night. When they entered into definitive REM sleep, they were promptly awakened and given a fifteen-minute vigilance task to complete before they went back to sleep. With this protocol, their REM sleep was selectively and significantly reduced. After these trials, their ability to tolerate pain (the heat from a lightbulb) was measured. The studies by Timothy Roehrs demonstrated that the REM-deprived subjects were more pain intolerant. Even more impressive was the fact that these effects could be seen after a relatively short period of REM sleep deprivation, potentially as little as a four-hour REM sleep deprivation. Beyond the contribution of sleep disturbances to acute pain perception, researchers have also linked sleep disturbances to the development of chronic pain conditions. In a 2015 study, rats who got insufficient sleep before an injury were more likely to suffer from chronic pain than well-rested rats.

There are all kinds of crazy things that happen to individuals as they enter into REM sleep, and knowing about them can give you

---

39 "Ma'am, I'm going to need you to wake up and push. The baby is almost here" ranks near the top of things rarely overheard in a labor and delivery ward.

some really fun things to talk about at parties. For example, humans are a euthermic species, meaning we are warm-blooded. We can, to some degree, regulate our body temperature in different environmental conditions. We can sweat when it's hot and shiver when it's cold. Animals like snakes are ectothermic (or poikilothermic) or cold-blooded and rely on environmental temperature to warm their bodies. That's why they have to hang out on warm rocks in the sun to get their body temperature up. It's interesting that you're no better than that copperhead when you dream at night because during REM sleep, you stop regulating your body temperature. Think about that. During dreaming, your brain completely suspends the fundamental and complicated function of temperature regulation.[40]

Light Sleep

Every great creation needs a solid foundation, and for our sleep at night, light sleep provides the bedrock for our night of dynamic sleep. Light sleep represents the state between being fully conscious and either being in deep sleep or dreaming. In light sleep, we are usually unconscious, but some people can retain a sort of awareness during this stage. It is usually fairly easy to awaken from this sleep as well, and so because of this, the state is relatively fragile.

Light sleep can be further subdivided into stage N1 sleep and stage N2 sleep. N1 sleep represents the transition from wakefulness to sleep. In a normal night of adult sleep, only about 5 percent of the night is spent in this stage. During N1 sleep, the brain waves begin to slow down and the darting eye movements that characterize wakefulness become slow and rolling. Muscle activity begins to diminish.

These changes continue in N2 sleep, which represents a deeper stage of light sleep.[41] Unique brain wave patterns called sleep spin-

---

40 Fellow partygoers will be much more interested in this fun fact than in how things between you and your in-laws have been a little strained lately.

41 This is deeper light sleep, not lighter deep sleep. Huge distinction.

dles and K-complexes are seen during N2 sleep, which help differentiate between N1 and N2 during a sleep study.

Almost half of an individual's night is spent in N2 sleep. Through N2 sleep, all other stages flow (see Figure 4.6). Diagnostically, this is important. If transitions to deep sleep and REM sleep are disrupted, the individual will spend more time than normal in stage N2 sleep. Because light sleep is not terribly restorative, these individuals will feel like their sleep is poor and unrestorative, and in some cases, they may feel as if they were not sleeping at all. Now you know why! This is crucial for those of you who think you never sleep to understand: You are sleeping, but you might be spending a disproportionate amount of the time in light sleep.

Deep Sleep

Deep sleep is the stage of sleep that seems to be the most poorly understood by patients. Apparently there is a group of grandparents out there telling their grandchildren the following pearls:

"Any sleep after midnight does not benefit your body."

"One hour of sleep before midnight is worth two hours after midnight."

While such advice is patently false, the origins of these bits of wisdom probably have everything to do with the function and timing of deep sleep, referred to by sleep specialists as N3 sleep.

Stage N3 sleep constitutes deep sleep. Deep sleep is sometimes called slow-wave sleep or delta sleep because of the slow brain waves seen during this stage of sleep (delta waves are the slowest of the electroencephalogram, EEG, waves). Older texts further divided deep sleep into two separate stages—stages 3 and 4. This division was based on the amount of slow waves seen during a thirty-second portion of sleep (called an epoch), with stage 4 having more slow-wave activity than stage 3. We do not subdivide deep sleep anymore. It's all just N3 now.

Typically, adults spend approximately 25 percent of their night in deep sleep with the majority of deep sleep activity occurring during the first half of the night. This sleep is restorative sleep and makes individuals feel rested (not sleepy) the following day. This is probably where grandma got her sleep advice.

Why is deep sleep restorative? Mainly because the time you spend in deep sleep happens to also be the time of greatest growth hormone (GH) production. I know, I know. . . . You've finished growing, so you are wondering why growth hormone is important. Basically it does so many things to help your body stay young and healthy, and perform better, it's a wonder people don't try to acquire it illegally and inject it into their buttocks in strange places like the locker rooms of professional sports teams.[42]

Forget the syringes! You don't need them. Simply make sure you value and protect your deep sleep at night and your brilliant brain will make all kinds of GH at night while you sleep, leading you to feel refreshed the following day. In addition, this GH will help strengthen your muscles and fortify your bones, help you recover from injuries and boost the functioning of your immune system.[43]

With all of this growth hormone floating around, we will be young and beautiful forever, right? Unfortunately, no. The amount of deep sleep declines as individuals age, and with it, so too does our GH secretion. Kids typically have tons of deep sleep. Ever been traveling back from the grandparents' house with the kids in the car? You stayed too late because you don't see them that often and you felt guilty having just gotten there yesterday, so here you are pulling into your driveway at 11:00 P.M. The kids are in their car seats, heads against the windows in a cute yet awkward position, sound asleep. So asleep, in fact, that you can lift them out of their chairs, take them to their rooms, strip them, throw on their pj's, brush their teeth, and they still don't wake up. Now that is some quality deep sleep.

As we mature, that GH gravy train dries up a bit as deep sleep lessens. This lack of deep sleep often makes individuals a bit more sleepy and/or struggle more with their sleep.

---

42 This is why I advise the athletes I work with to protect their sleep so that they can maximize their human growth hormone (hGH) production. This is essential for recovering from the wear and tear of their sport.

43 Ever wonder why staying up late studying for your exams always leads to illness and a C+?

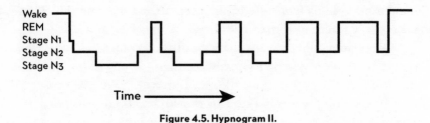

Figure 4.5. Hypnogram II.

Knowing this, how does it change the way you think about yourself when you fall asleep during your quarterly sales call? It should be a loud and clear signal that you are not receiving your daily allowance of deep sleep, and you are paying a big price for that oversight!

Sleep Cycles

The stages of sleep ebb and flow in a predictable pattern during healthy sleep. Complex chemical reactions in the brain initiate transitions from one stage to the next. By tracking the presence of these stages during a sleep study, a visual representation of sleep stages as they progress can be made. This is called a hypnogram (Figure 4.5).

Does the graph look familiar? It should because it's the second time it's popped up in this chapter. In this version, I've changed the names of the sleep cycles to their scientifically correct names, but otherwise, it's the same. Why put it in the book twice? Because it is essential for you to see what your sleep is doing or should be doing at night. This graph reinforces the concepts of progressively longer REM cycles as the night progresses and the majority of deep sleep occurring during the first half of the night. These concepts will help you understand why patients may exhibit unusual dreaming behaviors with sleep apnea, or why patients with insomnia often "wake up every night on the hour." Throughout this book, we will refer to these hypnograms to help you understand the patterns of sleep that underlay various sleep disorders.

Take a look at diagram 4.6. Unlike the hypnogram examples that show the specific timing of sleep cycle changes an individual

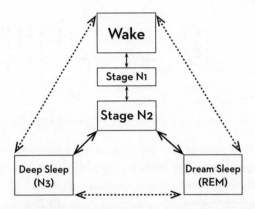

Figure 4.6. Sleep state transitions normally occur through stage N2 sleep. Transitions that bypass N2 sleep (dotted lines) are bad news!

undergoes during a typical night of sleep, this diagram is different. The diagram illustrates the normal pathways and abnormal pathways of sleep stage transition. Movement along the solid lines creates a normal night of sleep. Notice how sleep is not a direct march from wake to light sleep to deep sleep to dream sleep. This diagram really demonstrates the central role N2 sleep plays in the flow of normal sleep. Put your finger on the "Wake" box. Now move into stage 2 sleep (usually via stage N1). From there, one could go into deep sleep for a bit, back to light sleep, dream a little, back to light sleep, and then wake up without ever leaving the solid lines. As your finger moves along the solid lines, you are essentially creating a hypnogram, exactly like the examples we saw previously.

Now consider the abnormal dashed paths. Movement along these paths is considered abnormal. Imagine someone being awake and suddenly dropping off into a dream (REM sleep). This is a phenomenon called cataplexy, and it is abnormal. Here is how that might look on a hypnogram (Figure 4.7). See the drop from wake to REM at the onset?

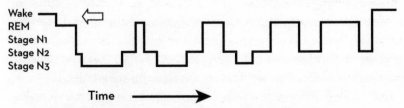

Figure 4.7. A patient going directly into REM sleep ... not cool!

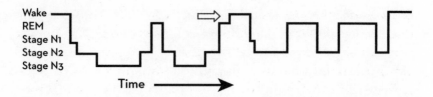

Figure 4.8. A patient going directly from REM sleep to wake ... also not cool!

Consider the reverse—a patient waking directly out of REM sleep as seen in figure 4.8. This is often the pattern seen in nightmares or sleep paralysis and is also abnormal.

By following the detours along the dotted lines, we can create many unusual or abnormal sleep stage transitions. Notice how the movement directly from deep sleep to dream sleep (and vice versa) is not normal. We will discuss these problems in greater depth later in the book. I just want you to have a feel for how these things are supposed to work.

One last bit of sleep advice you are likely to encounter has to do with sleep cycles, hypnograms, and how you can plan out your sleep to be more beautiful, healthy, and successful. The advice is based on the fact that we tend to sleep in cycles that last an average of ninety minutes. Life hackers have taken this information, consulted our grandparents, and come up with the sleep tip that we should sleep in ninety-minute cycles for optimal health. Some articles even suggest that the amount of sleep we need is irrelevant and it's only about making sure we time our sleep to end at a mark that is an interval of ninety minutes. This is what real scientists refer to as total horseshit (TH). TH usually consists of three key components:

1. Some grounding in science: ✓
2. Some blogs articles reporting amazing benefits from applying the TH: ✓
3. Zero scientific studies supporting the claim: double ✓✓

Keep in mind the ninety-minute cycle is an average. Maybe I have eighty-minute cycles. Maybe you have hundred-minute cycles. That's a big difference. Also, we usually have four to six cycles through the night, meaning that the time my third cycle ends could be an hour off from when your third cycle ends. Trying to time these things perfectly is a little absurd. As someone who looks at sleep studies all day, I can tell you that hypnograms follow a general pattern, but they are not precise. Also, does it really make sense to you after reading even just the early chapters in this book that the amount of sleep we need is unimportant? It's like saying the amount of food we eat is unimportant; it's all about ending the meal with a cream-based dessert that matters. People who follow the method of waking up only in increments of ninety minutes can often really miss out on a lot of sleep over time. Consider this example:

> John goes to bed every night at 11:00. He read an article online that said that sleeping in ninety-minute cycles could help him basically become Bradley Cooper in *Limitless,* so of course, John is all in. John needs to be up between 7:30 and 7:45 to make it to work. Unfortunately, neither of those times lands on a ninety-minute increment from 11:00 P.M., so he sets his alarm for 6:30, essentially robbing himself of at least one hour of sleep every night. So instead of getting eight to eight and a half hours of sleep, he limits himself to seven and a half hours.

To be clear, I have no issues with John getting seven and a half hours of sleep if that is what he needs. I do have an issue with him arbitrarily getting less just so he can awaken at 6:30 A.M. What happens if there is an early meeting and he has to be at work a little earlier? Does he now have to get up at 5:00 A.M.? Ridiculous. So if someone out there is already following this method and having great success in your life, good for you, but please spare me the email about it. I'm frankly as interested in your story as I am in hearing about your breakthrough with a psychic.

Some personal sleep monitors have the ability to awaken you during naturally occurring "lighter" stages of sleep, typically when the device detects that you are moving around a little. While there are no compelling studies supporting this practice as performance enhancing, it probably makes sense, as waking up out of REM (when the device detects that you are not moving—remember, you are paralyzed) can feel pretty awful. Probably the best way to avoid the need for such an alarm is to try to make your wake time as consistent as possible. For those readers who cannot (for example, a shift worker with rotating schedules), these kinds of alarms might be helpful.

....
CHAPTER 4 REVIEW

1. Your sleep is divided into three distinct stages—light sleep, deep sleep, and dream sleep.
2. Dream sleep is also known as REM sleep and is important for memory and mood regulation.
3. A lack of deep sleep can cause sleepiness because it is the most restorative phase of sleep.
4. These three stages in a healthy sleeper should flow in a predictable pattern.
5. Get all the sleep you need and work toward consistency in your sleep schedule. Consistency does not mean you must wake up at a ninety-minute interval.
6. Seek truth and avoid TH.

Damn, look at you now. In just four chapters, you've really come a long way. You now understand that while you don't feel like your sleep is working for you these days, you do sleep some. You have also made a determination as to how sleepy you are and how much or little and how well or poorly you must be sleeping. Finally, you are now gaining an appreciation for how sleep is structured and how it ideally should work (in other words, your goal sleep).

Can you achieve this goal sleep? Frankly, I'm not sure. Your sleep is a lot more screwed up than I initially thought when you started reading this book. I'm kidding! Of course you will! Chin up. Read on.

5

VIGILANCE AND AROUSAL
(Sorry but Not That Arousal)

WITH ALL THIS TALK OF sleep and how it works, it's a wonder you are able to stay awake and read this book. What possible black magic is allowing you to fight the forces of slumber just one little chemical cascade away from erupting in your brain?[44]

Vigilance. *Vigilance* (sometimes termed *arousal*) is the medical term we sleep doctors use to describe the wakefulness-promoting systems in your brain that allow you, in most cases, to decide when you will be awake. For some, it's not working well enough. "Officer, can you let your accident report reflect the fact that my client's car is wrapped around that telephone pole because of inadequate vigilance?" For others it's working too well. "Yes, Chuck Norris, please tell me more about the Total Gym. I don't care if it's 3:00 A.M. and I have to be at work by 6:00." Vigilance can be your best friend when your sleep works, and it can be your worst enemy when it does not.

Vigilance is a rapidly changing entity. Imagine sitting in a meeting that has gone on forty-five minutes too long. You see the speaker's lips moving, but you're thinking about the weekend or what

---

44 Remember adenosine and melatonin?

items you need to pick up from the store on the way home. You might even be fighting the urge to close your eyes if you are comfortable enough. Suddenly, you snap back to reality when your boss stops his presentation to sarcastically ask if he can get you a pillow or blanket. The room is quiet and all eyes are on you. You frantically wonder if you fell asleep as you wipe away some drool from your lip. *Presto!* You are now vigilant. You are wide awake, breathing quickly and well aware of your pulse pounding in your ear. You're feeling many things right now, but sleepiness isn't one of them. How can it be that a split second ago, you were literally falling asleep in front of your coworkers yet now you don't feel one ounce of sleepiness in your body? Vigilance.

Vigilance does not just happen when your boss catches you sleeping through a meeting. It can happen anywhere. Open up a cabinet and see a mouse. Suspenseful ending to a movie. Shopping, eating, fire alarms, watching a nail-biter basketball game—all kinds of places. Any event or activity will increase vigilance if it captures your attention.

Every yin has its yang and in the world of sleep, the flip side to vigilance or arousal is sleepiness, or how likely you are to fall asleep. Fortunately after reading Chapter 3, you're an expert on sleepiness now.

As vigilance is reduced, the likelihood of sleep initiation increases. Conversely, as the drive to sleep (sleepiness) is reduced, the likelihood one will become vigilant increases. The presence or absence of vigilance will determine if that wakefulness is sustained or not. That shouldn't be surprising. Wake up at night to a quiet, dark house and your quietly sleeping spouse, and as you roll over, vigilance is low and you go back to sleep, sometimes not even remembering you woke up. Wake up at night next to a grinning clown with tangled red hair and enormous shoes and sleep is suddenly nowhere in your immediate future.

The processes in our brain that control sleepiness are distinct from those that control wakefulness. This is an important concept. For many centuries, sleep was merely viewed as the absence of wakefulness. In other words, the theory was that there was one process, a single variable—a light switch, so to speak. When you were awake (switch on), the brain's wakefulness was high. Sleep

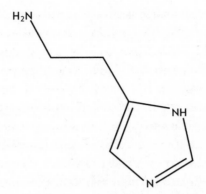

Figure 5.1 Am I the only person who thinks histamine looks like a sperm?

was the light switch of wakefulness being turned off, like the brain was a lightbulb. When the individual was asleep, the brain was turned off. One variable: on or off.

So understanding that there are chemical underpinnings for sleep (like adenosine, melatonin), what are some examples of chemicals responsible for wakefulness? I'll bet you already know them; you just don't know that you know them.

The first chemical to be aware of is histamine. Histamine is a chemical that produces wakefulness in our brains. Knowing this, you can imagine the effect of a drug that blocks histamine. These "anti" histamine drugs would make us sleepy, which is exactly what this class of drugs does, in addition to helping us with our allergies and motion sickness.

CUTTING-EDGE SCIENCE

A 2015 STUDY PUBLISHED IN the *Journal of the American Medical Association* (*JAMA*) looked at medications that block the chemical acetylcholine. These anticholinergic medications have been strongly linked to the development of Alzheimer's disease, a condition that centers around a lack of acetylcholine. Because many antihistamines are also anticholinergic as well, these drugs (like Benadryl) were included in the study. The study concluded that the long-term cumulative use of these medications was associated with an increased risk of developing dementia.

The take-home message is this: Have an occasional runny nose after mowing the grass? No problem. Are you someone who takes an antihistamine every night to help you sleep? Stop now. The drug is unnecessary and may lead to long-term problems with your memory and cognition.

Incidentally, along with the first-generation antihistamines, antimuscarinic medications like oxybutynin for overactive bladder and tricyclic antidepressants like Elavil (amitriptyline) were included in the study and were also linked to dementia. I have seen many patients who are taking oxybutynin for overactive bladders at night to help them sleep (often erroneously, as it is really their sleep apnea and not their bladder that is the problem) as well as amitriptyline to help them fall asleep. In other words, it is not inconceivable that there are people out there taking multiple anticholinergic medications on a nightly basis to help them sleep better. If I'm describing the contents of your medicine cabinet, you should give your primary doctor a ring.

Another important chemical for wakefulness is dopamine. Dopamine does all kinds of things in our bodies. Because dopamine is the chemical that is lacking in Parkinson's disease, it is easy to see how important dopamine is for smooth, coordinated movement. Dopamine is also the neurotransmitter of pleasure, so whenever we do something fun, our brain gets a little hit of dopamine. That's okay when it comes to sex and chocolate bars, but it can be not such a good thing when it comes to addictive or destructive types of behaviors.[45]

Outside motivation, movement, and reward, dopamine plays a central role in terms of our wakefulness. This is why Grandpa, who has a little Parkinson's, is always falling asleep. He's missing a key chemical for staying awake. Suffice it to say I see many patients with Parkinson's because this lack of dopamine really affects their sleep in a negative way. These individuals are prone to developing REM behavior disorder, a condition in which the normal paralysis of REM sleep is impaired, leaving the individual free to act out her dreams. Parkinson's patients often have difficulty with restless legs

---

45 Unfortunately for some, the urge to get that dopamine surge is inescapable and leads to addiction. Dopamine is usually the central player in addictive behaviors.

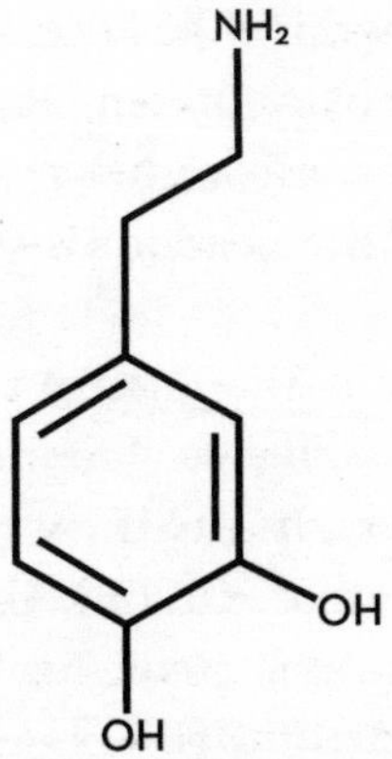

Figure 5.2 This is the reason why you can't eat just one Lay's potato chip.

and frequent limb movements at night, while struggling with extreme daytime sleepiness. This sleepiness and resultant sleep often creates tragically unpredictable sleep schedules.

One more wakefulness chemical I'll mention is a relatively new discovery. Orexin (or hypocretin[46]) is a chemical central to wakefulness. Funny story: It was discovered and named by two different

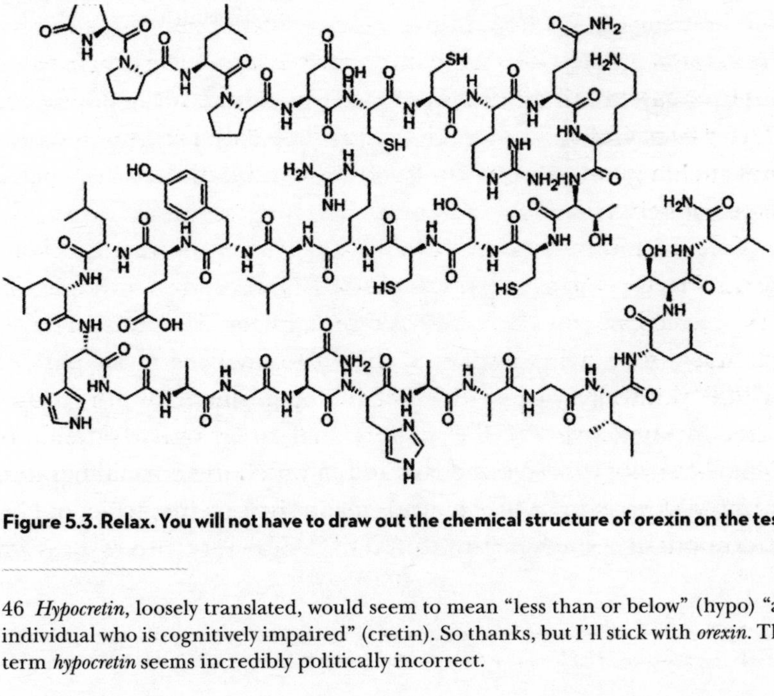

Figure 5.3. Relax. You will not have to draw out the chemical structure of orexin on the test.

---

46 *Hypocretin*, loosely translated, would seem to mean "less than or below" (hypo) "an individual who is cognitively impaired" (cretin). So thanks, but I'll stick with *orexin*. The term *hypocretin* seems incredibly politically incorrect.

labs, so now the same chemical has two different names. As they say on *Iron Chef*, "Let the battle begin." And begin it has, as scientists and scholars have battled over which term to use. I went with the definitive resource, Wikipedia, and used orexin because hypocretin has been reduced to a "redirected from" status.

The absence of orexin is the cause of a condition called narcolepsy. Narcolepsy causes extreme sleepiness, which makes sense because there is not enough orexin to go around. We will talk more about narcolepsy and orexin (as well as dopamine) in Chapter 15. I just wanted you to put these chemicals in the proper context of wake-promoting chemicals.

So on Team Sleepiness, think about adenosine and melatonin. On Team Wakefulness (or vigilance or arousal), think histamine, dopamine, and orexin. Think of these teams as being two distinct systems making up two opposing forces, as shown in the drawing below.

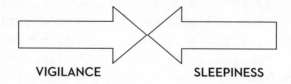

VIGILANCE   SLEEPINESS

This model describes states of human arousal or sleepiness.

A normal individual awakens in the morning with a baseline amount of vigilance and a low level of sleepiness because the prior night's sleep reduced the adenosine levels in the brain.

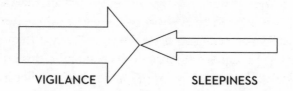

VIGILANCE   SLEEPINESS

As the day progresses and that individual works hard at her job, swings by the gym at lunch, and heads back to the office in the afternoon, sleepiness begins to grow.

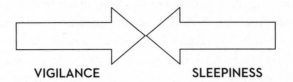

VIGILANCE   SLEEPINESS

Depending on the circumstances of the day, vigilance may be diminished (in a long meeting, driving on a boring stretch of highway, and so on). When sleepiness gains too much of an upper hand on vigilance, sleep occurs.

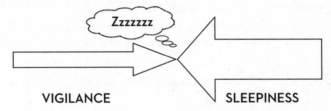

VIGILANCE   SLEEPINESS

This is an example of excessive daytime sleepiness (EDS); look around—it's everywhere. Anytime there is unwanted sleep during the day, that's EDS. With more vigilance, most people can hobble through their day without falling asleep behind the wheel. Sometimes it means driving in 28°F weather with the windows down and the air-conditioning on, singing along to the radio and chain-eating chili cheese Fritos with a Mello Yello chaser.

Eventually, even the most bizarre rituals cannot overcome the awesome power of sleepiness. As you can see by the model above, anytime sleepiness gets the upper hand over vigilance, sleep will occur. This is a good thing at night, so even when you've been sleeping well, sleep can still overtake your brain at the end of the day. After an inadequate amount of sleep at night, this model illustrates how sleep can gain the upper hand much earlier in the day.

For some, it's the exact opposite problem. Their day of work has produced exactly the same amount of sleepiness during the day, maybe more. They work harder during the day, they exercise harder at the gym, and they come home from work more exhausted in the evening. Sadly, when they get into bed, they "can't shut their brain off and sleep." Pity . . . all of that accumulated adenosine going to waste.

How can this happen? It would be unusual for an individual stranded and starving to death on a desert island to refuse the first meal offered to them upon their rescue. Even if the food offered was not their favorite, it's most likely going to be consumed voraciously. So what possible force could be preventing this hard worker from getting into bed and going to sleep?

Everybody sleeps. You know that. The longer an individual is awake, the bigger that sleepiness arrow gets. Unless that individual sleeps, that arrow gets nothing but bigger.

Wake up.

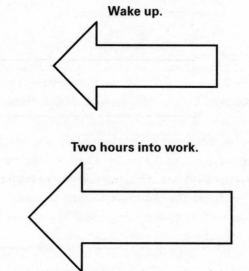

Two hours into work.

After lunch (circadian factors often contribute to this as well).

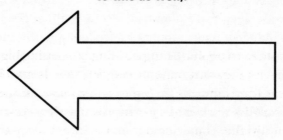

Quittin' time (a small reduction in sleepiness as circadian factors rush in to help us make it to bedtime).

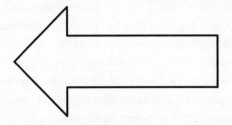

Watching shows about tiny houses and other reality nonsense starts boring us.

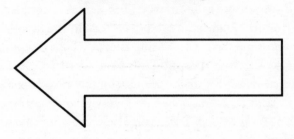

Staying up so late, programming has stopped and the infomercials begin.

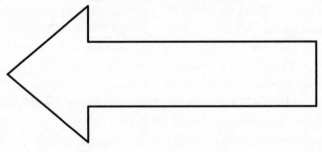

Look at that arrow. It is doing nothing but getting bigger. Sleepiness does not wax and wane during the day. It marches relentlessly upward and onward. *The longer you are awake, the more adenosine you are accumulating and the more driven your body is to sleep.*

Many individuals describe not being able to sleep if they miss their golden sleep time. "I am usually able to fall asleep around 10:00 or 11:00. If I miss that time, forget it. I'll never get my brain to shut off."

Stop and think about this statement. Now think about that ballooning arrow of sleepiness. As the day goes on, it is never getting smaller, so that when 10:00 P.M. rolls around, you might be ready to sleep. If so, good night. If not, try again in another hour. As that time passes, you will only become sleepier!

Why would somebody feel that despite their busy day, they can't seem to fall asleep in their bed? I often hear tales of patients struggling to stay awake and watch the late news, but as soon as they get up to walk into their bedroom and go to sleep, they feel wide awake. What's happening?

Vigilance is happening as that person suddenly experiences arousal. Commonly, it's temporary, but usually enough to cause a little frustration. "Damn, I was so sleepy out in the den—where did that sleepiness go? Why can't I fall asleep?" That sleepiness didn't go anywhere. Vigilance has suddenly been increased, just as if a fire alarm had gone off. From this arousal state typically come frustration, anger, and resentment, which feed vigilance and lessen one's chance of falling asleep. Suddenly there is stress that you aren't going to sleep and you begin to think along these lines: "Do I have any Benadryl I could take?[47] Maybe I'll take some of those pills my spouse takes. Maybe if I'm not asleep within an hour, I'll get up and cut a Xanax or Klonopin in half and take that. Here, I'll start my stopwatch to keep track of this. Maybe I'll put the TV on the timer and see what's on. I wonder if I have enough sick time to call in tomorrow. I wonder if astronauts have trouble falling asleep. Why am I thinking about astronauts? Oh God, I'm never going to fall asleep . . . what's wrong with me? Why do I feel so cold inside? My mother never loved me. . . ."

This patient is clearly *not* going to sleep anytime soon. What's more is that every night that this goes on, the individual begins to develop more dread toward the act of going to sleep. Benign things within the bedroom often trigger these feelings: "Here's my bed. There are the pictures on the wall of me on that Segway tour of Cancún. Yep, this is the room where my sleep sucks." It's no accident that many people with insomnia sleep *better* in hotels or when they visit other people. Those same cues that are reminders

---

47 I can't remember where I put it. What's going on with my memory?

of your track record of terrible sleep aren't there. In fact, I often have patients share the fact that their sleep is much better in the guest room compared with how they sleep in their own room.

These cues create pressure. Pressure to go to sleep. That pressure creates vigilance and the inability to initiate sleep. It's like an insomnia-training program the individual has been engaged in. Come home; fix dinner; eat; watch TV; start dreading sleep; start worrying about how many sleeping pills you have left; start resenting your spouse, family, and friends who sleep well; and start cranking up the anxiety about your sleep. So as that sleepiness arrow grows, its growth is being dwarfed by the rapid growth of that vigilance arrow that is quickly increasing in size at a time of the day when it should be dwindling.

 VIGILANCE EXERCISE

THINK ABOUT YOUR FRIENDS. NOW think about the friend who always talks about what a wonderful sleeper she is—you know, the woman who just can't seem to understand why you have so many sleep problems when she just drifts right off when she gets into bed. You know her: She's healthy, happy, and productive. Works hard during the day, sleeps well at night. Great abs, no cellulite, boobs like a junior in college (you'd never know she's the mother of a four-year-old and a seven-year-old!). You hate her.

Now that you've chosen your great sleeper, you are all set for your Vigilance Exercise.

For this exercise you need a printer. (Cutting out individual letters from magazines would work, but it's very labor intensive.) Now, create a letter for your friend following the template below. Don't sign it! It will ruin the funny surprise!

wE have abDucted your cat. Do not call the police. We are watching your eveRy move even now. Our demands are simpLe. Once You get into bed tonight, iF you do not faLL asleep within the subseQuent 4 h0urs, you will nEver sEe ApolLo Creed agAin.

Observe and record what happens to your friend's sleep from the bushes outside her home.

This exercise is an absurd suggestion, but think about it for a moment. This woman is not going to be successful in falling asleep within four hours. She's sleepy, and she has years of good sleep under her belt, but the anxiety she will feel starting with the reading of that note will all but guarantee her failure at the task. What has changed? This individual is now *working* to fall asleep rather than allowing sleep to simply happen. It's like the proverbial kid on Christmas Eve who knows Santa is coming to fill his living room with presents to be unwrapped in the morning. He just needs to go to sleep first . . . but he can't.

This phenomenon is seen in places outside of sleep. Steve Blass was a pitcher for the Pittsburgh Pirates during the late 1960s and the early 1970s. While he was a great clutch pitcher, Blass will always be remembered for suddenly losing his control of the ball. This happens periodically during Major League Baseball seasons. I remember Steve Sax of the Los Angeles Dodgers, a second baseman, suddenly losing his ability to throw the ball accurately to first base. Despite years of making this throw over and over again, it suddenly happens. What's worse, the more focused and stressed out one is about the sudden malady, the worse it becomes.

Blass's coach was famous for instructing Steve to "try easier" rather than the "try harder" mantra we have all heard from our coaches. Overthinking, stress, and anxiety can ruin activities that we consider automatic. Blass had no difficulty throwing strikes during warm-ups, but in the game, he couldn't find his target to save his life. Most people with insomnia caused by increased vigilance sleep in a similar way. When they get into bed at 11:00, they can't sleep to save their life. However, if they are watching the news in the La-Z-Boy when they get home from work, they often nod off immediately. What's the difference as between the La-Z-Boy and the bed? The same difference as between the warm-up pitcher's mound in the bull pen and the pitcher's mound in the center of the sold-out stadium.

Vigilance and anxiety can be important reactions. Without them, we couldn't awaken to the smell of smoke at night and react to save ourselves and our family. We would simply be too sleepy. Anxiety makes the world go around. I want my president to worry about things. I want my accountant to be an anxious guy. I don't want my surgeon to be a laid-back individual. I want her to be one tense bundle of nerves.

So there are forces at play that make us sleepy, and there are forces at play that wake us up. Imbalances in these forces result in sleep problems. In these cases, it is essential that we focus on reducing the anxiety surrounding the act of sleeping. This book was, in part, born out of a desire to help my patients reduce the anxiety that often surrounds sleep by providing them with knowledge about sleep.

So what exactly is your sleep problem? Too much sleepiness? Too much wakefulness? Do you even know? It's difficult to get a handle on what's happening with your sleep at night because you are in fact sleeping. How many hours are you sleeping? Before you answer that question, read on. Your ability to answer might be strongly influenced by the very subject of this upcoming chapter.

CHAPTER 5 REVIEW

1. Vigilance or arousal counteracts sleepiness and works to keep us awake.
2. This can be a positive thing, or if vigilance is too great, it can be a real problem.
3. Throughout the day, the balance between sleepiness and vigilance changes. That's what wakes us up in the morning, and drives us to bed at night.
4. Steve Blass won a hundred games as a Pirate and kicked serious ass in the 1971 World Series, allowing only seven hits and two runs in the eighteen innings pitched. He was the runner-up to the great Roberto

Clemente for MVP of the World Series that year. His throwing issues did not define him.

Your sleep issues do not define you either. Keep reading and learn why some of the things you believe to be happening when you sleep may simply not be true.

6

SLEEP STATE MISPERCEPTION

How Did This Drool Get on My Shirt?

ONE OF THE FIRST PATIENTS I evaluated in my private practice came with the urgent complaint that she had not slept for the last six months. When this highly anxious woman said she had not slept, she didn't mean she hadn't slept much; she meant at all, and she was scary serious.

You know now that this is impossible, but she did not. To begin working on a patient's sleep problem, the patient and I need to arrive at an understanding that everyone sleeps some. Sure, humans have the ability to pull the occasional all-nighter and some highly motivated individuals have been able to push the limits of sleep deprivation in highly artificial circumstances. But other than that, we all sleep. I sleep and this problematic woman sitting in my office, staring at me and waiting for me to hand her the magic sleeping pills, sleeps too.

"Well, if I sleep, how come I see the clock change all night long? I watch TV all night long or sometimes I get up to iron."

"Well," I replied, "you probably do wake up and see the clock and notice things on TV, but you are squeezing some light sleep in there."

"How do you know what I'm doing? You don't sleep with me."

Indeed I do not. Things were heading down a dark road. Confronting people about their sleep when they think they are not sleeping is occasionally unpleasant.

Quick story. Once, I and my wife, Ames, went to see *The Usual Suspects*. The Atlanta theater was practically empty when we sat down. The film started out with a dark scene of shady characters running around and shooting one another on a boat docked in a harbor. Before the scene was over, Ames was out cold (for good reason: she was a schoolteacher at the time—the hardest job in the world). About an hour later, she was awakened by a loud noise; I'm going to guess more gunshots. She immediately said, "This movie is too dark and slow." In her mind, she had closed her eyes for just a second, almost as if she traveled through time an hour or so. She complained the rest of the time about the fact that the movie did not make any sense, yet she'd missed a critical hour of the story without realizing it. Weeks later I overheard her telling someone how bad the movie was. It annoyed me because I loved that movie. But she had no conception of how her sleep had altered her experience of seeing the film. She'd missed it, and in her mind, it didn't happen. She couldn't recognize her misperceptions.

And neither could my patient. She similarly could not recognize that she could look at the clock and notice the time, then fall asleep. When she wakes up and looks at the clock again, she assumes it's just the first time she's looked again. She doesn't recognize she has been asleep in between looking. In some cases, patients can even dream about looking at the clock and other mundane things that happen at night, yet not be able to distinguish their dream activities from reality.

Going back to my patient . . . Because we couldn't arrive at an agreement that she slept even a little bit, we scheduled her for an overnight sleep study as a way to measure and record her sleep in a more scientific way. With the sleep study, we would, among other things, be able to pinpoint exactly how much she slept by analyzing her brain and neurological activity during the night. When I read her sleep study, I saw that not only had she slept but she slept like a drunk fraternity brother.

When I saw her again to review the results of her study, her first words to me as I walked into the exam room were "Told you so."

"Told me what?" I asked.

"Told you that I don't sleep. How can anybody sleep with all those wires glued to your head and people watching you? I've still got glue in my hair."

"You not only slept but you slept a lot." At this point I produced a summary of her night's sleep, which came to six hours and forty-seven minutes, and in anticipation of her skepticism, I produced a video of her sleep. When I showed her the results of her study, she stood up with hellfire in her eyes. At this point she turned to her demure husband and griped, as if I were suddenly not in the room, "We're leaving. I told you he was too young to be a doctor," and she stomped out of the office.

What this woman experienced goes by many names, the most current of which is paradoxical insomnia. Paradoxical insomnia is the phenomenon in which an individual thinks she is either not sleeping or is sleeping a very reduced amount compared to her actual sleep, which often is fairly normal. In the past, it was referred to as sleep state misperception. Before that, it was called twilight sleep.[48]

When you think about sleep, particularly your own, you absolutely must forget everything you know or think you know about sleep in general and your own sleep. We are constantly being bombarded with misinformation about sleep that does more harm than good. For example, many people who feel like they don't sleep at all at night actually sleep a perfectly normal amount. Conversely, many individuals who feel they sleep great at night, but are tired during the day, are not in fact sleeping great . . . I am one of those people. Just ask my frazzled wife after she removes her earplugs.

One of my favorite questions to ask a patient is what his snoring sounds like at night. The fact that many of them attempt to answer the question illustrates the central problem of acquiring information about an individual's sleep: *He can't tell you about it because he is asleep.* Funny how this doesn't stop my patients from providing long and detailed explanations about how they sleep, their sleeping behaviors, and the neurochemistry behind their sleep. I once

---

48 This is an even cooler concept if you've seen the *Twilight* movies and know that Edward and his vampire clan do not sleep. As a sleep doctor, I can't support that reckless behavior, so I'm totally Team Jacob.

had a patient who prefaced her sleep complaint by telling me in a very matter-of-fact tone that her pineal gland had "disintegrated." As a reminder, the pineal gland is a small structure in the brain that produces melatonin (that sleep-promoting chemical, remember) in response to light. The woman had no evidence that this had occurred, no MRI pictures of her brain, no history of traumatic injury . . . nothing. It just made sense to her over time that her sleep complaints fit best with that explanation, so she adopted it. In the end, there was really nothing wrong with her at all besides some minor difficulties initiating sleep that she had escalated in her mind to catastrophic proportions.

When people think about sleep, there is a certain degree of artistic freedom that seems to be allowed. I've yet to meet an individual with a broken leg who would explain the event by saying that the metabolic processes involved with calcium regulation had gone awry and this led to the fracture. Most people simply say they fell and heard a pop. Although in many ways sleep is not much more complex than that, we make it out to be.

 ## SLEEP STATE MISPERCEPTION EXERCISE

1. Get married.
2. At night, watch some television with your soul mate.
3. Keep watching television until the person who shares a single beating heart with you closes his or her eyes and falls asleep.
4. Look at the clock; record the time.
5. When the love of your life awakens, look at the clock again. Record the time.
6. Ask your beloved how long he or she slept for. Compare it to how long he or she actually slept.
7. When Thanksgiving rolls around, you can do the same thing with your extended family as individuals fall asleep during the Detroit game.

The point of this exercise is simple: The reality of the time we sleep is often quite different from the perception of how long we have slept. Many of us tend to radically underestimate this time. It is common for individuals who are a bit anxious and who sleep lightly to experience this type of sleep. The important take-home message is that if you are reading this book and feel like you are not sleeping, you are not alone. In fact, you are so not alone that a sleep doctor devoted a whole chapter to this one phenomenon. It is also important to understand that even though these individuals are sleeping, that lack of sleep perception is not normal. I repeat: *Not feeling like you sleep, even though you do, is not normal!*

Beyond being abnormal, paradoxical insomnia can be a real soul-crusher. People like to sleep, and they get very disturbed when their sleep is not happening properly. While paradoxical insomnia is usually a primary condition, there have been cases of obstructive sleep apnea that presented like paradoxical insomnia, as noted in a 2010 study. There have been reported cases of individuals being so distraught and helpless because of their perceived lack of sleep that practitioners have resorted to electroconvulsive therapy to help patients "feel" their sleep.[49]

When it comes to sleep, everyone is entitled to feel that he slept. In other words, it is not my intention to simply prove to you that you are sleeping when you don't feel like you are and then leave you feeling that way forever. No way! Everyone is entitled to feel that wonderful amnesia that sleep brings. You hop in bed, kissy-kissy good night if there is someone else in bed with you, set your alarm, and turn out the light. What follows should feel like you hopped in a time machine that transports you to the sound of an alarm clock waking you for the day. That is the goal, and we can achieve it.

For many, simply relaxing and understanding that they really are not in danger of not sleeping cures their sleep woes. For others, it's tougher. Just use this chapter to help understand and describe your problem on a deeper level and realize that you might be getting more sleep than you realize.

---

49 Shocking, I know.

CUTTING-EDGE SCIENCE

In 2015, M. R. Ghadami published a study looking at the sleep of thirty-two veterans with documented PTSD and sleep-related difficulties. The individuals in this hyperaroused group reported an average of four hours and twelve minutes of sleep but actually averaged seven hours and six minutes of sleep. They estimated their sleep efficiency to be 59.3 percent (meaning that of the time they were in bed, almost 60 percent was spent in sleep), when in reality, their measured sleep efficiency was 81.2 percent. In addition, the test subjects estimated it took them an average of about seventy-six minutes to fall asleep when in reality, it took only about twenty minutes.

This study illustrates why as many as 80 percent of patients with PTSD suffer from paradoxical insomnia, and just how much of a role hyperarousal plays in our ability to perceive sleep. In fact, for many people who struggle with their sleep, you can almost think of their nightly angst as a mini-PTSD episode!

So if you're that woman who said I was too young to be a doctor: I'm older now and getting gray hair. While I still stand behind what I said about the fact that you sleep, I would still love to be able to finish what we started and help you feel sleep more than you do right now.

CHAPTER 6 REVIEW

1. It is possible to sleep at night and have a limited ability to perceive your sleep.
2. While not perceiving your sleep is not the same as not sleeping, it is still abnormal!
3. Start opening your mind to the possibility that this might be happening to you as you struggle to stay asleep "all night long."

You're asleep, you're awake, you're awake but really asleep, you're asleep but really awake. (I do this to get out of cooking breakfast.) It's so complicated. How does your brain keep track of when all of this should be happening? Read on to find out how the brain keeps you and your sleep on schedule. Hint: Here comes the sun!

CIRCADIAN RHYTHMS
The Clock That Needs No Winding

N 2007, THE NEW ENGLAND Patriots football team was accused of illegally videotaping the signals of their opponent during the first game of the season. The Patriots were caught and punished (largely because they were taping a team coached by a former Patriots coach). When news of this episode, dubbed Spygate, broke, it was revealed that this was not the first time this had happened and that the Patriots had been caught and warned about this behavior previously.

The reaction many had to this news was "Why would a team who had already been caught doing something illegal risk doing it again?" The answer is simple. It's a lot easier for a team to succeed if it can anticipate its opponent's next move rather than simply react to it.

Your body is no different. It likes to have a heads-up before you eat or engage in physical activity. The ability of your body to anticipate a large cheeseburger with fries and a shake is crucial for successful digestion.

How does your body accomplish this? Circadian rhythms. They govern virtually everything our bodies do. In the last chapter, I introduced the circadian system, but it deserves a chapter all to itself. Circadian (Latin for *circa*, "about" + a *dian*, "day") rhythms are internal processes within us that cycle about every twenty-four hours. They're pretty awesome and they require very little from us in terms of calibration—like a fancy watch that runs on body movement.

These rhythms are not only in humans, but are in virtually all animals, plants, and even fungi. To say these mechanisms are highly preserved would be a tremendous understatement! Why would you and a poppy need a circadian rhythm? The answer can be seen in the work of Jean-Jacques d'Ortous de Mairan. In a classic study, de Mairan showed that the heliotrope (touch-me-not flower) opened and closed with the sun during the day and that it also retained the ability to open even when kept in the dark. In other words, the plant has an internal ability to anticipate its environment (the movement of the sun) rather than simply react to its movement.

Evolutionarily, species that could anticipate their environment were ultimately more successful than those who couldn't. So fast-forward a few million years and here we are eating smoothies with wheatgrass extract and watching mixed martial arts on our flat-screen plasma TVs.

So if we don't depend on the movement of the sun to photosynthesize something to eat, why do we still need the sun? *Do* we still need the sun? About eighty years ago, two guys tried to answer that very question by going on the ultimate male-bonding adventure. Nathaniel Kleitman, the godfather of modern sleep medicine in this country, and his scientific partner, Bruce Richardson, left Chicago and traveled to Mammoth Cave in Kentucky, where they set out to "retrain" their circadian rhythm from twenty-four hours to twenty-eight hours. Their rationale was that if they could artificially adopt this twenty-eight-hour cycle and make it stick, it would prove that the human circadian rhythm was not internally driven, but simply a response to the environmental twenty-four-hour light cycle.

Kleitman and Richardson stayed in that godforsaken cold, damp cave for thirty-two days. While their experience was not on the level of *Midnight Express*, it was difficult. After a month, they emerged from the cave to a media storm of publicity (not to the degree we see today where celebrity couples have their names combined into one cute name, like Bruthaniel; but still, it was a lot of national attention for its time). Alas, after all of that, their findings were inconclusive, and since reality TV was years away, they never made a penny from this unusual stunt. They confirmed that individuals do in fact have an inherent rhythm that is slightly longer than twenty-four hours (twenty-four hours and eleven minutes but who's counting?). The period of time of an organism's circadian rhythm is often given the symbol τ. So in humans, τ = twenty-four hours plus change.

How convenient, since a day is approximately twenty-four hours long. What makes up that slight difference between the environmental time and our internal circadian rhythm? Our brain is able to get clues about the true external time and make little daily "corrections" to our own body clock.

A great way to think about this is with the cheap watch analogy. Imagine you and your friends all go out and buy cheap watches. Perhaps your watch runs ten minutes fast every day. Your friend has a watch that runs ten minutes slow. You put the watches on, and off you go about your lives.

If neither of you ever adjusted your watches, you'd find problems slowly start to creep into your lives, particularly your friend who has the slow watch. While day after day, you'd be arriving at appointments more and more early, your buddy would be getting places more and more late. On day one, he'd be ten minutes late. On day two, twenty minutes late. In less than one week, he'd be punching in at work over an hour late, picking his kids up from school over an hour late, getting dinner on the table over an hour late, and so on.

You'd be in a better situation for a little while. You'd be getting high marks for your promptness to the workplace and your cooking speed. Your kids would love you since they would get to leave school early. Eventually, though, even you would have problems. The dinner is cold since it has been sitting around; your boss would suspect you were brownnosing for his position.

Finally you come to the conclusion that your watch is crap, but you decide to try to make it work. You decide that every morning, you are going to turn on the *Today Show*, reset your watch by their clock, and tackle the day. And guess what. It works. In fact, the more you check with the *Today Show*, or any other broadcast that provides the time of day, the more accurate you are with the timing of your day.

In this example, several things are illustrated. First, it's better to have a watch that's a little fast than a little slow if we are trying to anticipate things (and not be late). That's why human circadian rhythms are twenty-four hours and eleven minutes and not exactly twenty-four.

In addition, we need time cues like the sun to help set our internal clock every day. These cues are referred to as zeitgebers, and the sun is probably the strongest of the bunch.

Other zeitgebers are mealtimes, exercise, social interactions, temperature, and sleep. These clues to our external time happen frequently and give our bodies cues as to how to adjust its internal time. The more zeitgebers an individual is exposed to, particularly cues that are presented at uniform times every day, the more synchronized an individual's circadian rhythm.

These minute daily adjustments often go along without a hitch unless there are sudden or dramatic changes in these zeitgebers. These sudden changes are most commonly seen in jet lag and shift work.

With jet lag, the environmental time cues are suddenly altered, and its effect depends on the direction of travel and number of time zones crossed. This movement produces many unpleasant symptoms, including sleepiness and difficulty sleeping, digestive problems, reduced motivation and impaired concentration. These symptoms make perfect sense if we think about those cheap watches.

If you travel from Atlanta to Las Vegas, you are essentially suddenly changing your external time by three hours, but your internal circadian rhythm is initially unaffected. In other words, as you walk into the Bellagio, your brain is still on Eastern Standard Time. This shift produces all kinds of problems as you sit down to eat dinner in the hotel. The heavy cream pasta dish and strawberry cheesecake are entering your digestive system at 10:00 P.M. Vegas time, but your brain hasn't gotten that memo yet. It thinks it's 1:00 A.M. and therefore is wondering why Alfredo sauce has suddenly found its way into your gut at a time when you should have been asleep for two hours and squarely in REM sleep. You can imagine how prepared your digestive system is—it's not! Suddenly, your stomach is reacting to the $41 entrée rather than anticipating it.

And your ability to have a conversation that night at dinner is gone. All your mind can think of is "go to bed." You're sleepy and not thinking clearly, which pleases the management of the casino. It's going to be okay. With every day you spend on the Strip, your body is able to adjust by one time zone that you crossed. Since you crossed three time zones, you'll be rolling sevens in as little as three days . . . just in time to be getting on a plane for home, feeling much lighter without all of that money crowding your wallet.

Jet lag affects many people, but you don't have to be a world traveler to experience the mental fog and loose stools of the jet-set crowd. If it's the thrill of falling asleep in meetings you crave, run—don't walk—to the nearest shift work job and get yourself hired.

Shift workers are everywhere. They are driving that large eighteen-wheel truck that shares the road with you. They are taking care of your loved ones in hospitals across this country. They are the pilots and flight staff that flew you home from Vegas (pilots get the double whammy of having a shift-work job that features jet lag—*jackpot!*).

In shift work, the environmental cues remain constant while your schedule varies (in jet lag, the cues change in your constant environment). Compared to jet lag, the effects of shift work can be equally hard on the various systems of your body. Factor in individuals who work jobs with rapidly changing shifts (three graveyard shifts, followed by two morning shifts) or people with two jobs, and you have the fixings for trouble.

When you ask many doctors to think of a condition that produces truly pathological sleepiness, many think of a patient with narcolepsy. Patients with narcolepsy are often overcome with sleepiness and can suddenly drop off into dream sleep. Perhaps some would picture an overweight individual with severe sleep apnea. Good guesses, but if you want to get serious about sleepy, you need go no further than the shift worker. In 2001 the diagnosis "shift work sleep disorder" came into being (the name has since been shortened to shift work disorder). Now, for the first time, the sleepiness associated with this type of work could be looked at officially as an illness.

An illness, just because someone's working the graveyard? Give me a break. We're Americans. We can handle it. Just give us some strong coffee, and we're good to go.

Not quite.

One way we can measure an individual's sleepiness is a simple test called a Multiple Sleep Latency Test (MSLT). In this test, an individual is allowed to sleep one normal night and is awakened the next morning. At that point, the patient is allowed to do anything she wants except sleep for the next two hours. When those two hours are up, it's nap time. The patient is put back into bed and given the opportunity to take a brief nap. Oops, not too long. After a few minutes, the patient is awakened (if she fell asleep), and once again, she is invited to stay up for another two hours with another nap opportunity to follow. This pattern continues until sometime in the afternoon, typically after five naps. Afterward, we can look at the amount of time it took the individual to fall asleep (if she slept within the five twenty-minute opportunities).

Patients with narcolepsy and sleep apnea are often sleepy, but they are usually much less sleepy than individuals with shift work disorder.

 SHIFT WORK EXERCISE

1. This sleep exercise requires dice. Go find some.

2. Roll one die.

3. Look at the result, and go to bed at the time indicated:

   ⚀ = go to bed at 10:00 P.M.

   ⚁ = go to bed at 2:00 A.M.

   ⚂ = go to bed at 6:00 A.M.

   ⚃ = go to bed at 10 A.M.

   ⚄ = go to bed at 2:00 P.M.

   ⚅ = go to bed at 6:00 P.M.

4. Repeat steps 1 to 3 every night for a month. Keep track of how you feel from day to day.

Notice how some nights you are getting into bed unable to sleep. Other nights it's a struggle to stay awake until it's time for bed. Shift workers who work schedules that rapidly change have it even worse. Shift workers, on average, lose six hours of sleep per week when compared to their non-shift-working counterparts. It is a tough existence.

 ## CUTTING-EDGE SCIENCE

TREATING THE CONSEQUENCES OF SHIFT work can be difficult, and like your family at Thanksgiving, everyone has her opinions and those opinions are often only loosely based in reality. To shed some light on this area of sleep, Juha Liira, a researcher at the Finnish Institute of Occupational Health, set out in 2015 to look at the use of medications in the treatment of shift work disorder and determine if they really helped. The study found that melatonin provided an average of twenty-four minutes more daytime sleep when used in conjunction with night shift work, but it did not help workers fall asleep faster. The study also found that stimulants like modafinil and armodafinil provided workers with increased alertness. It is interesting that hypnotics like zolpidem did not seem to lead to any improvements in sleep quality or performance. This was one of the first reviews of how we currently treat shift work disorder. We are still looking for the best way to help people who have to live with shift work.

CHAPTER 7 REVIEW

1. Circadian rhythms dictate everything we do, including when we get sleepy and when we feel awake.
2. It is important that we consider our mealtimes, exercise, and light exposure when it comes to trying to establish a healthy circadian rhythm.

3. Jet lag and shift work represent examples of circadian
    rhythm disorders.

Wow. You've made it to the intermission. Go stretch . . . make
yourself a cup of Earl Grey and relax for a while. Let what you've
just read settle in your brain. Those first seven chapters were heavy.
They deserve some time for reflection and contemplation.

That's great. Now when you're ready . . . let's attack your sleep
issues head-on. No longer hampered by misinformation, fear, and
quasi-mythology, you're an educated sleep ass-kicker now. Nothing
can stop you from a good night's sleep.

Intermission

YOUR CERTIFICATE OF HIGHER SLEEP education is in the mail. Display it proudly, preferably over the headboard of your bed. Look on it confidently when you go to sleep every night as a reminder that you know what's really going on when you sleep.

Now, what to do with this knowledge? My guess is that you have some current issues with your sleep you want to fix. Outstanding! Knowing what you know now, getting your sleep straightened out should be a snap.

Let's look at some different sleep difficulties. Sleep problems can usually be divided into two major groups: those that make us feel that we don't sleep enough and those that make us feel like we are sleepy too much.

As a sleep doctor, I believe that everyone who walks through my door is essentially in one of these camps. Let's look at these groups a little closer and see how our knowledge about sleep helps us understand what might be going on behind the scenes in each of these exclusive sleep problem clubs!

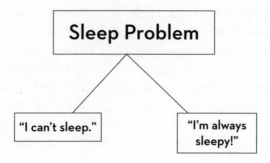

"I Can't Sleep"

Now I know and you know (because you've read this book) that members of this group are sleeping, but there is something going on keeping them from being satisfied with their sleep. Okay, so what could that be? For this answer we need to look at the individual and the environment in which he sleeps. For individuals who have a sleep environment that is not conducive to sleep, they need to clean that up or establish better sleep hygiene. I'll show them how in Chapter 8.

For many people, establishing better sleep hygiene is not enough for them to find the fountain of great sleep. For them, difficulty sleeping or feeling as if they cannot sleep is a huge problem. In Chapter 9, we will dive headfirst into the strange and misunderstood world of insomnia. Insomnia is something that most of us will deal with transiently from time to time. Trouble falling asleep, trouble staying asleep, waking up before your alarm goes off, and being unable to return to sleep are all examples of the insomnia people deal with, and Chapter 9 will help you put these problems behind you.

For others, it becomes a prison cell from which they feel escape is impossible. In Chapter 10, this chronic insomnia, or what I refer to as "hard insomnia," is addressed.

Wrapped up within the world of "I can't sleep" is the sleeping pill, which people mistakenly consider the easy fix. Sleeping pill use has risen to a culture within this country. There is a huge group of individuals who feel as if they need sleeping pills to fall asleep. Sleeping medication origins, current practices, and dangers are discussed in Chapter 11.

Chapter 12 takes a step back to look at the individual's sleep schedule. For many people who feel they cannot sleep, their problem lies not within their ability to fall asleep, but rather in their unrealistic expectation of how much sleep they need. For many, simply understanding how to properly establish a better sleep schedule can help them find better sleep.

Chapter 12 also looks at the flip side of seeking too much sleep: scheduling an inadequate time for sleep. Thus this chapter serves as our transition to the world of the excessively sleepy because many people who are struggling to stay awake are doing so because they are not allotting themselves enough time to sleep. Individuals who struggle to get out of bed in the morning may find their problems with sleepiness are rooted in their schedules. We will also take a closer look at shift workers and the extreme sleepiness they are forced to deal with on a daily (or nightly) basis. And so into the land of the bobbing heads we go.

"I'm Too Sleepy"

Within the too-sleepy group, we start with one of the single biggest indicators that an individual might not be sleeping enough or may be struggling with the quality of his sleep: the nap. Chapter 13 looks at napping: how it can be a healthy and effective tool for your sleep but also how it can be working against you.

Next, in Chapter 14, we focus on the group that makes up the lion's share of sleepy patients: sleep apnea patients and their friends the snorers.

Chapter 15 deals with other diagnoses that cause excessive daytime sleepiness, conditions ranging from restless legs syndrome to narcolepsy.

Finally, Chapter 16 addresses the sleep study and who should sign up for one.

To make things easier, I've provided a visual layout of the remaining chapters and how they fit together.

One final note about the second half of the book. From time to time, I will make recommendations about various products or devices that might be beneficial to your sleep. It is important for you

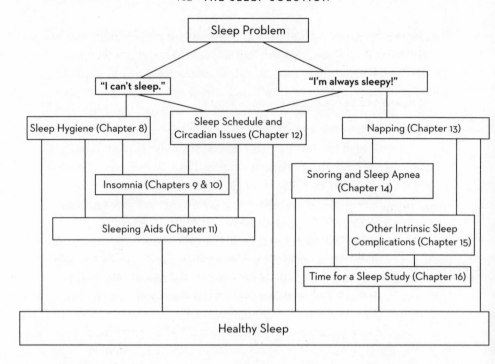

to understand that while these devices are not necessary for sleep, they can enhance sleep for some people. Think of it this way: Tortilla chips are good. For me, adding salt and a hint of lime takes them to a new level. Don't get me wrong; if unsalted and un-limed chips are available, I'm going to eat them. I would never say, "I cannot consume bland chips," particularly if I am hungry. However, that spice addition really makes it a better experience all around.

 VACATION CRUISE SLEEP EXERCISE

IMAGINE YOU ARE HEADED OFF to go on a relaxing Caribbean cruise. You arrive at the port in sunny Miami and board the ship, *Slumber of the South Seas*. It's beautiful, the drinks are paid for, and your upgraded room has a deck overlooking the ocean.

You return to your stateroom after a fantastic dinner and a show only to find that you have forgotten to pack your _____ [white noise machine, teddy bear, sleep mask, blue-blocker sunglasses, inflatable companion that vaguely resembles Sofía Vergara in dim lighting]. You can't remember the last time you slept without it (or her).

Choose the best ending for your story:

1. As you get into bed, you think, "No big deal" and after a few more minutes than normal, drift off into a relaxing sleep.
2. You start to panic as you try to think of a place where you could buy or steal your forgotten item. Anxiety grips you tightly as you realize the chances of you sleeping on this trip are slim to none. You feel yourself begin to hyperventilate as you make the decision to storm the bridge of the vessel in a desperate attempt to take over the boat and return home.

Your answer needs to be 1. Feeling dependent on your sleeping pill, noise machine, special blanket or continually running radio is not a position you want to be in. Break the habit.

See these products for what they are: They are minor assistances, small enhancements. Can you sleep well without them? Probably, but these things might help to make your sleep move from good to great.

On with the show.

8

SLEEP HYGIENE
Clean Bed Equals Sleepyhead

NOW YOU HAVE THE KNOWLEDGE—THE book learnin', so to speak—about sleep that you need. Basic training is over. Here comes the hard part. Are you ready to test this knowledge on the battlefield of flowery comforters and useless throw pillows? Figuring out you've got a kidney stone is easy. Fixing that problem can be quite painful. Well, fixing your sleep problem is similar. Expect things to get a bit worse before they get better.

I chose sleep hygiene as the starting-off point for the second part of this book because that's exactly what sleep hygiene is in terms of fixing your sleep: a starting point. Chances are it is not going to solve your problems, but if it does, fantastic! Look at all of the time you'll save not having to read the rest of this book. But also don't be discouraged if sleep hygiene alone doesn't solve all your sleep problems. Sleep hygiene is the foundation needed to fix all sleep problems, but it's not uncommon that on its own it doesn't completely solve the problem.

Sleep hygiene is the act of controlling your sleep behaviors and environments in an effort to optimize your sleep. Basically, it's doing what you can to set yourself up to sleep successfully. It is controlling what you can control.

Many patients I treat know a thing or two about sleep hygiene.

It's a subject that is discussed everywhere . . . self-help books, morning news shows, sleep websites.

Patients often say things like "I've tried everything. I don't watch TV in bed, I don't exercise late at night, and I never drink coffee after lunchtime." And they still cannot sleep.

What they need to understand is that sleep hygiene is like cleaning up your place before you throw a big party. You have to sweep, tidy up the place, maybe buy some new candles or something. My, the place looks great. Does all of your work mean your Dungeons & Dragons party is going to be a huge social hit? No, of course not, because Dungeons & Dragons is probably not a great foundation for a memorable house party. It doesn't matter how perfect the setting is. The basis of the party is flawed, so the whole thing is going down in flames.

Your sleep is no different. Getting things tidy and in order before you sleep is very important and can be the basis of some minor sleep disturbances. Remember those rats with dirty cages? There are many ways to get your own cage in order and most are pretty obvious.

Creating a Hibernation Lair

There are many steps that go into transforming your bedroom into a sleep lair, and step one involves light. Make your bedroom dark: really dark. Remember how melatonin makes you sleepy but only if your eyes aren't seeing the light? Well, block the light, every bit of it, if you want to sleep well.

Make your bedroom like the guest bedroom in my parents' house. When I was in fifth grade, my folks finished the basement in our house and created a bedroom that did not quite meet code, as it had no windows. It's surrounded by earth on two sides and is nowhere near an exit. In other words, in the event of a fire, it's a death vault. It's so dark in that room, it's hard to find your way out in the middle of the day! And because of this, that suburban catacomb is the best damn place to sleep in southwest Virginia. If nobody ventures down to find me, I could easily sleep until hunger awakens me. Believe me, ridding your room of light can help.

And when I say make your room dark, I mean *dark*. Your brain is like the zombies on *The Walking Dead*—it can pick up on the smallest sources of light such as a clock radio, a cell phone display, the crack underneath your door. So turn off your phone (or better yet, keep it in the kitchen) and turn your clock away or do something to block the light it emits. You don't need to know what time it is at 3:15 A.M.[50]

A big source of light is the television. Why did televisions wind up in the bedroom? I don't have a clue. To me, it's like a toilet in the family room. Televisions produce loads of light. This high degree of light intensity in addition to the noise and stress they produce can significantly worsen sleep. Also, the television is conditioning you to need it to fall asleep. Not good. Watch your TV somewhere else.

Every time I give a lecture, I ask the audience, "How many people sleep with their television on all night long?" I would estimate that about one out of every twenty-five people I speak to admits to their television droning on all night long, flooding that bedroom with light and noise. "It relaxes me," or "I like a little background noise" are common excuses for the television being on all night.

When it comes to your lair, dark and quiet are best, and that flat-screen is ruining both. Still think that wall-mounted Sony does not affect your sleep? Consider the 2014 study showing that subjects continue to mentally sort words from spoken word lists even after they fall asleep. Can you learn Spanish in your sleep while you hypnotize yourself to lose weight? No. Does that mean there is not a light on in your brain once you fall asleep? Nope. Remember our theme: The brain is doing amazing things when we sleep. Turn the TV off. Your brain does not need to listen to infomercials and *M*A*S*H* episodes all night long.

Ridding your bedroom of the faintest light is important for everyone, but doing so is particularly important if you are someone who must sleep during the day because of shift work or other unusual work schedules. If light simply cannot be eliminated, buy a soft contoured eye mask to prevent light from reaching your eyes.

---

50 Oh damn, it's 3:15. I'll never get back to sleep now that I possess this information. Oh, how I long for the ignorance of not knowing what time it is early in the morning.

Buy a few, find one you like and make sure you put an extra one in your travel bag. (This is one sleep aid you really don't want to do without. If you're on that cruise with no sleep mask, use a towel. In a real pinch, use your arm to block the light. We all need to sleep in the dark.)

 SLEEP CAVE EXERCISE

1. Go into your bedroom, close the blinds, shut the door and turn off the lights.
2. Put your hands in front of your face. Can you see them?

If yes, keep working. Your room isn't dark enough.

If no, congratulations. You can turn the lights back on.

Aha! How did you find the light switch to turn the lights back on if your bedroom was a sanctum of absolute darkness in which no hint of light could exist? Stop telling me what you think I want to hear and eliminate that light source.

One last point I need to make clear, and then I promise I won't belabor it throughout the rest of the book: turn off your smartphone, laptop, tablet, or any other electronic device. Completely off. That light is killing your sleep. In a 2014 study by Charles Czeisler, individuals who used e-readers before sleep at night took an average of ten minutes more to fall asleep and had less REM sleep than individuals who read a printed book with indirect light. Any light exposure in the late evening or early night can have a negative impact on your circadian rhythm and sleep, so keep your environment dim at the end of your day for great sleep. If you must use light, try either to filter out the blue and green from your device or consider wearing blue-blocking glasses. Screens and similar lights should be turned off several hours before going to bed.

PRODUCT SUGGESTION

IF YOU SIMPLY MUST HAVE your laptop on at night, consider installing f.lux, Dimmer, or a similar app that will reduce both the amount of light and the quality of the light you are exposed to. These apps work well, both are free, and neither will bring a bunch of unwanted junk onto your computer. Apple's Night Shift mode works in a similar way on the iPhone.

Want to go slightly less high tech? Buy some Uvex blue-blocker glasses. These glasses block the blue/green lights emitted by the screens and lights in your life. Pop them on whenever you can't escape light at night to help with your sleep initiation later in the evening. Bonus: These glasses will make you look and feel like Bono, which is never a bad thing.

Get Cozy in Your Crib

Along with light, your bed should be comfortable. Patients frequently ask me which mattresses they should use. The simple answer to that one is "I don't know." Everybody has different comfort levels on different mattresses. That's why there are so many different kinds! I like firm, but others like soft. Some cultures sleep in hammocks. Batman sleeps hanging upside down. The main thing is be comfortable and don't get suckered into buying a bed as a cure for all of your sleep woes. If your bed is hideously uncomfortable, upgrading might be quite helpful, but be smart with your money. Comfort, that's all you need.

A few more things while we are talking about beds. No sense in having this nice bed and covering it in uncomfortable bedding. Invest in your bedding. Buy some high-thread-count sheets and a snug cashmere blanket or a goose down comforter. Again, we all like different things, but you should *love* your bed. Buy the nicest bedding you can afford.

Because of ambient temperature differences, variations in our body temperature set points, metabolic and exertional levels, and medications, some people feel hotter at night than others. In addition, night owls often feel warmer later at night than morning types. If you are someone who is extra cold at night, try flannel sheets. If you are someone who gets hot at night, you might want to try sheets that wick away moisture and sweat. Again, it's not going to solve your problem, but it's essential you feel comfortable in bed.

 PRODUCT SUGGESTION

DEEPSPORT IS A COMPANY THAT produces bedding that has some unique properties. First, the bedding is very cool to the touch, so if you are someone who sleeps hot or sweats, this can be quite helpful. In addition, the material reduces bacteria, allergens, and organisms like dust mites and bedbugs from getting into the fabric. My personal favorite product is their "sleep sack." It's a sleeping bag–like product made from their material. It is easy to travel with so it can provide a clean and cool sleep environment away from home. Better yet, if you use it at home too, when you travel your brain will feel like it never left your own cozy bed, which might help you sleep.

SHEEX is another company that has been around for a while making similar performance bedding. They also make sleepwear made out of the same moisture-wicking material.

For the ultimate in bedding temperature control, invest in a ChiliPad. As God is my witness, this water-cooled mattress cover will change the way you sleep forever. The device pumps warm or cold water (you decide) through little tubes in your bed all night long, allowing you to sleep at your perfect temperature. You can even buy one that has a zone for you and one for your partner. My bed has come to be known affectionately as Korea. In North Korea (my wife's side), it's a hot mess, literally. She likes her bed warm when she gets in, and then she promptly turns her side off. My side, South Korea, is kept at the coolest setting all night long, and it is glorious. My bed feels amazing on my old legs, particularly after exercise that day. My side of the bed is so cold, I'm pretty sure if I kept lunch meat in there, it would not spoil.

While it is probably totally obvious, I'm often shocked by how many people have put no thought into their pillow selection. It is as if people just grow up with a certain pillow on their bed, they bring it with them when they head off to college, it follows them when they move in with their partner, and nine pillowcases later, it's still there on their bed.

Do you even like your pillow? Is it comfortable? You do understand that you have made no formal commitment to this pillow, right? Get out there, find a new one to sleep with, sleep with several at a time and keep the one you like best. Some pillows are made with shredded latex that allows the user to add or remove the shredded latex from the pillowcase so that the pillow is the exact firmness you prefer. If, over time, the pillow is getting a little too flattened, more can be added. Memory foam pillows often provide excellent orthopedic support of the neck and spine, but like their mattress counterparts, they often trap heat so individuals sensitive to being hot at night may not like them. Feather and down pillows tend to be very light and soft. They are washable and breathable but may flatten over time. Patients with allergies can struggle with these pillows. Even if you don't have allergies, you might be annoyed by being poked by the feather variety. Other options include wool, cotton, buckwheat, and synthetic materials like polyester. Take some time to find what works best for you.

With your bed outfitted with comfortable material, you should probably do the same thing for your body. Dress comfortably. I typically encourage people to wear less, not more to bed. If you are someone who gets cold at night, you can always drag on blankets or more covers during the night. Wearing a fleece-lined flannel jumpsuit to bed can be problematic if you start to get hot in your sleep.

Last but not least, find a quiet clock (no ticking!) that does not light up your room, set your alarm, and forget about the time. It's best if you can't see what time it is when your room is dark. It makes no difference what time you wake up and pee in the night, but for many, that little fact is enough to set off a full-blown anxiety attack. Spare yourself. If you wake up before your alarm goes off, just remind yourself that it's still time to sleep. That's the only thing that matters.

The den of dormancy is looking great. You've got your new sheets and comforter, perfect pillows, and a dark and quiet clock, and you're in the right pajamas. You're so pleased with your work that some positive anticipation is starting to creep into your mind. Fantastic. A big hurdle in the sleep of many is ridding themselves of the negative feelings they associate with their bedroom.

Don't underestimate these feelings. Imagine a child living in a house with an abusive father. Every day, when the father gets home from work, he yells for the boy to come downstairs. The father then proceeds to verbally unload all of his day's stress and angst onto the boy in a vicious verbal tirade. It happens every day at the same place at the bottom of the stairs. Fast-forward in time. The boy has long since escaped the unhealthy home. He's happy and well adjusted. (Okay, he had a little therapy.) He has a family of his own and does not carry the same problems of his father, who died years ago. How do you think this grown-up son feels when he walks through the door of his old home during a family visit with his elderly mother? Even though years have passed, as soon as he walks through the door to that landing where he was yelled at years ago, those same feelings come rushing back as if it were yesterday.

Your bedroom may be that landing. This is why change in the bedroom can be helpful. Sure, the mattress change, the new blinds, and the cozy new comforter serve a practical purpose. They also serve an equally valuable role of changing your surroundings so that your bedroom is transformed into a new haven—a place that your mind does not readily recognize as "The Place Where My Sleep Sucks." Keep this in mind. Some of you might want to not limit your bedroom makeover to the bed. Throw some new paint on the wall. President Harry Truman believed, as others believe, that a soothing blue gray is best for sleep. My own bedroom walls are blue gray. Avoid canary yellow or bright reds, as they can be arousing. Buy some new artwork; change things up. Out with the insomnia cell, in with the Suite o' Snooze.

CUTTING-EDGE SCIENCE

IF YOU WANT TO TAKE your sleep to the next level, consider the relationship between sleep and nature. A 2015 *Preventive Medicine* study showed that individuals, and in particular men, seemed to sleep better when they were exposed to green spaces and nature.

Be creative with your bedroom setup. If you can come up with ways to foster more of a connection with nature (for example, try sleeping on your screened porch when the weather permits or eat your meals outdoors every day), you may find your sleep is improved.

Strange Bedfellows

Your room is done and it looks great. You love it and for once, you're actually looking forward to sleep. The big question now is: who else is invited to the party? Do you live alone? The answer is easy then. The only variable is you. Have a spouse and seven pets? The situation is a bit trickier and may involve fleas.

Some spouses are great sleepers, like me. I turn down the covers for my wife every night, scatter rose petals about and set little choc-olate mints on her pillow. I never move in my sleep, I give her all the covers she wants and I am totally silent. I also have the unusual ability to massage her feet during the night as I sleep. I'm very much aware that as bed partners go, I am the exception.[51]

For many (I'm speaking from clinical experience here and not personally, of course), spouses can be a real problem for their part-ner's sleep. Their partner's snoring is described as the sound of small animals being executed. They pull all of the covers off of you at night, leaving you shivering in the fetal position. They kick and fling their arms about, sometimes inadvertently striking you and

---

51 Really, you don't need to ask my wife.

leaving mystery bruises all over. They talk and moan and get up to pee so often that you can't establish any kind of sleep rhythm. Their alarms are set to get up early and the noise they make in this process essentially means you get up at that time too. Does this sound familiar? Wait. Have you been talking to my wife?

Cheer up. Nowhere in those vows did you say that you had to sleep in the same bed with that person every night.[52] I'm going to get into some topics here that might not sit well with you (or your partner) at first, but hear me out. In fact, why don't the two of you read this section together . . . holding hands? Great. Look into each other's eyes. Tell each other how much you love one another. Read on.

Every sleep doctor worth his or her weight in foam earplugs will tell you the bed is for two things: sex and sleep. I'll say it again: The bed is for sex and sleep. It is not for watching television. Save that for the living room, as the living room is for two things as well: watching television and sex. Notice that sleep is not on the living room activity list. Typing on the computer, talking on the phone and paying bills are all no-no's in the bedroom. Sleep and sex. That's it. By that rule, if your partner is preventing your sleep from happening in the bedroom (he reads with a bright light; she snores), there's a problem. If your sleep isn't happening because your partner is trying to have sex with you, you might need to work on communicating more effectively.

Now, let's focus on sleep.[53] The conversation here is a little more complicated. If your bed partner is an impediment to your ability to sleep, something needs to be done. Here are the options:

1. Do nothing and ignore the problem, which will only worsen over time, leaving you tired, irritable, and harboring deep-seated resentment and contempt for your partner and his or her annoying sleep habits.

---

52 Just avoid the beds of others mainly.

53 One more thing about sex and sleep. Sex can help promote sleep through several mechanisms. First, sex is a physical activity, which, as you know, increases sleep-promoting adenosine. Sex often happens in the dark, promoting melatonin secretion. In addition, an orgasm promotes prolactin release, which acts to suppress wakefulness-promoting dopamine in the brain. Finally, sex produces oxytocin in our brain, which facilitates relaxation and positive feelings, helping us relax and sleep.

2. Convince your partner to seek help for his or her snoring, leg-kicking, teeth-grinding, hair-pulling, talking, moaning, screaming, dream enactment, choking, or whatever behavior is waking or keeping you up.
3. Sleep in separate bedrooms. Sleeping in separate locations can be subdivided into:
   - Permanent separate locations
   - "As needed," when one partner leaves if there's trouble
   - Scheduled time apart. For example, sleeping apart every Tuesday and Thursday with the other days co-sleeping. This sometime happens naturally when one partner travels. In my case, when I'm on call or getting up early to exercise, I sleep in the guest room so my early-morning rumblings and pager do not disturb my wife.

Obviously, I do not condone doing nothing (option 1). Some nocturnal behaviors can indicate significant problems that can carry with them serious health risks in addition to the toll they are taking on your sleep. Option 2 is almost always my pick; if you are able to convince your partner to seek help for his or her condition, I highly recommend it. This is the best route to take. If doing nothing is not an option and you are unable to convince that stubborn spouse to bring this up with a doctor, you're stuck with option 3.

It's important to communicate effectively here, or feelings can get hurt quickly. I believe that everyone has a right to sleep well. If only one canteen is taken on a hike, most couples would share the water inside. It would be unthinkable for one partner to drink it all, leaving the other with nothing. Why should sleep be any different? Why should one partner be allowed to deprive another of something that in many ways is just as important as water or food?

Sleeping together in a bed is a powerful symbol of marriage, unity, and love. Not sleeping together, in the minds of some, is an act of separation or lack of commitment. I spend a lot of time in my clinic telling couples it is okay to sleep apart sometimes. I have dubbed this a "sleepcation." It costs nothing but is restorative in the same way a vacation can recharge one's batteries. Sometimes, an individual may simply need time to get his sleep back on track. At that point he may be able to rejoin his partner. For others, it may be a more permanent arrangement. Sometimes, this separation can enhance the partner's enthusiasm to seek help.

If there is an openness to sleeping apart, there is no right or wrong way to do it. For some, starting out in the same bed and then moving to a different room when the lights go out works well. Again, choosing specific days of the week can be helpful in terms of eliminating guilt. "It's Tuesday, so I'm sleeping in the guest room tonight." Having specific days to sleep apart prevents the individual sleeping away from feeling like there is a decision to be made every night at bedtime as to where she will be sleeping. Sometimes a trial period of sleeping apart can be useful to see if, in fact, the partner's sleep habits are to blame for your poor sleep.

Virtually everything I've mentioned (except the sex stuff) can also apply to pets in the bed. In my opinion, when it comes to your bed, there is no room for pets. If you have a pet that sleeps with you and your sleep is great, fine. Rover stays. If you are having problems with your sleep, and you have a tiny suspicion that your pet is to blame, the pet really should go.[54]

The snoring spouse is dealt with, the dog is in the basement, but there may be one or more individuals who still need to be jettisoned from your bed. That's right, your children.

Family beds are always a hot button of controversy, so brace yourself. I am against them. I'm against them not only to protect your own sleep but because helping your kids establish the ability to sleep consistently and confidently is invaluable. This means they must be able to initiate sleep by themselves and with minimal crutches (rocking, pacifiers, blankets, night-lights, and so on). If

---

54 Interesting sidenote about pets: Many patients in my clinic who go on to be diagnosed with obstructive sleep apnea syndrome, a condition in which they stop breathing at night, often complain about the fact that their dog wakes them up during the night. For most, the owner reports the event as highly annoying, being awakened from a peaceful sleep by licking. Later, once the individual is diagnosed with sleep apnea and treated, the dog stops waking him. I've dubbed this phenomenon the "Lassie effect," as it is my belief that the dog is aware of the choking and breathing disruption and is simply trying to get its owner to breathe. To date, I've never spoken to a cat owner who's described this occurrence. My theory is that, like the dog, the cat is aware of its owner's breathing distress, but it doesn't do anything about it.

that little peanut sleeping next to you is affecting your sleep, keeping this arrangement is not doing you or him any favors. Time for him—or God forbid, *them*—to be introduced to his own bed.

I'm going to be blunt because I have seen someone very close to me deal with this: co-sleeping can be dangerous. It does not take much for a child to be smothered by an adult. If you think it cannot happen in your family, you are wrong.

Your Dirty Habits Are Keeping You Awake

Despite their esteemed place within popular culture, when it comes to habits that cause significant problems with sleep, it would be hard to find behaviors worse than drinking alcohol or smoking cigarettes.

What you need to know is this: Nicotine is a stimulant. It will keep you awake and worsen the quality of your sleep once you fall asleep. The exact nicotine amounts don't really matter, just like the exact caffeine amounts in beverages don't matter that much either. Stop smoking, especially anywhere near the time you are going to sleep. If you smoke in bed, stop. It's not only exceptionally bad for your sleep but it's also a really dangerous place to be smoking. (Seriously, in 2005, 24 percent of Canadian smokers reported falling asleep while smoking in the year before the study. This is why I never sleep during trips to Canada.) I'm not a smoking expert. If you smoke, ask your doctor, family, and friends to help you quit. There are much better and cheaper bad habits out there that won't hurt your sleep and drain your wallet. Try knuckle cracking or nail biting.

Caffeine is a stimulant, so guess what I have to say about it. It's not helping your sleep. Cut down on it, particularly in the evening near your bedtime. It keeps you awake and makes you pee. In 2013, sleep researcher Tom Roth performed a study showing that caffeine consumed as much as six hours before sleep could reduce sleep time by as much as one hour![55] Tea and chocolate have sim-

---

55 Keep in mind that it is likely that these individuals may not perceive the significant sleep loss that caffeine creates. In other words, if you are someone who believes that the coffee you drink after dinner or later at night does not affect your sleep, you may be wrong.

ilar properties. So if you are struggling with your sleep and you have a barista-grade espresso machine in your home, it might be time to stop drinking coffee or at least cut back on your cups of joe, especially within six hours of going to bed. Is it hard to reduce or stop your consumption of these products? Sure, but you can do it. Wean yourself slowly or go the cold turkey route. You're tough. If you work at Starbucks, ouch! You better pack your lunch and a caffeine-free beverage.

Alcohol is hell on your sleep too. It worsens the quality of your sleep, leads to awakenings at night (often to pee), worsens breathing problems at night including snoring and choking (apneas), and may magically produce surprise bed partners in the morning. Think of it this way: As with most things we consume to "help us sleep," alcohol produces sedation, not necessarily sleep. With all of this bad publicity, it's amazing that alcohol remains the number one sleep aid in this country. Why is alcohol so popular? There are probably several answers to that question:

1. You do not need a prescription for alcohol. I get it. Nobody wants to go see a doctor. It's expensive. It's time consuming. Our waiting room magazines are so old, they have pictures advertising the third Harry Potter movie. Having to wait around to get a doctor to not only approve a sleeping pill but write you a prescription is no picnic. Contrast that with alcohol, which is easy. All you need is some cash and an ID (or a cool older brother), and you can purchase enough spirits to launch you into oblivion tonight.

2. Alcohol is sedating. Many people associate a good night of sleep with rapid loss of consciousness once they decide to retire. Alcohol can provide that. Alcohol tends to make individuals fall asleep faster, but this effect does not tend to translate into either more sleep or, more important, improved performance the following day. In other words, is the quick loss of consciousness better for you in the end than staying up a little later to read your favorite book? It most certainly is not.

3. Alcohol promotes amnesia. Another criterion many people have for achieving a good night of sleep is having no memory of what happens between falling asleep and waking up. Alcohol can help with this. This is sometimes referred to as a blackout, and it is not going to help you knock your presentation out of the park tomorrow morning. You would probably be better off staying up all night!

It should be noted that some studies report an increase in deep sleep associated with alcohol, typically during the first half of the night. While this is disputed, what's not disputed is the mess alcohol makes of the second half of your night as the alcohol is metabolized. Ever wake up about four to six hours after a binge and find it absolutely impossible to go back to sleep? It's like an amazing Caribbean cruise that ends with your boat sinking. Is there anything in the first half of the cruise that could compensate for the soggy ending? When it comes to alcohol and sleep, don't fall into this trap. If you are using alcohol to help you sleep, stop. If you have a problem with alcohol, get help.

I know sleep doctors tend to belabor their warnings against nicotine, caffeine, and alcohol. I'm not going to waste your time or get into an argument with you on this one. Nicotine, caffeine, and alcohol worsen your sleep.

"But I like my double grande mocha-latte."

ZIP IT!

"It's just one cigarette before . . ."

Seriously, are we really having this argument?

"Come on, you can't be suggesting I give up my Shiraz with . . ."

Ugh. If you are having trouble with your sleep, these behaviors need to be modified or stopped. People commonly ask questions like "How many cigarettes can I smoke without affecting my sleep?" or "Is my morning coffee okay?" or "Are my two glasses of wine at night okay for me to sleep?" Unfortunately, there are no clear answers to these questions that have any rigorous scientific backing, but that's okay. We can still approach these questions intelligently. If your sleep is positively lovely both in terms of how you feel about it and how you feel during the day (not sleepy), then your glass of wine at night is probably okay. Keep in mind, what you have come to view as being great sleep may not be great, so if you are feeling adventurous and experimental, pick a two-week period and avoid your glass of red during this time. Pay attention to your sleep quality and the way you feel at the office. If you wear a Fitbit, or similar device, take a look at your average sleep quality measures before and after the wine experiment. If you see no change, the wine is probably fine. If you feel better not having it, it's really up to you to decide if the change is important enough for you to choose water with dinner!

Diet and Sleep

Now that your coffee and wine are history, what other items can we remove from your Trader Joe's cart that will improve your sleep while at the same time making your nighttime eating dull and unappealing?

Well, when it comes to food, the National Sleep Foundation feels it best not to eat any of it within two to three hours of bedtime. While there is no definitive research as to precisely how long to wait between food and sleep, this is probably a good number and should help you avoid the sleep disturbances some people feel from indigestion or gastroesophageal reflux if they go to bed too soon after eating. Foods heavy in protein can have the unwanted effect of keeping you up at night.

If you simply must munch on something in the evening, think about Thanksgiving. Ever notice how incredibly sleepy you feel after your big meal? People always blame the tryptophan in the turkey, but in reality it's the carbohydrate bomb you dropped in your stomach that does the trick. Eating that massive amount of sweet potatoes with sugar-coated walnuts, stuffing, cranberry sauce, and pecan pie creates a sudden surge in your blood sugar and a spike in insulin levels that promote a feeling of sleepiness. In 2007, research by Chin Moi Chow from the University of Sydney showed conclusively that a high-glycemic-index meal consumed four hours before sleep resulted in a significantly shorter time to fall asleep than a low-glycemic-index meal.

Take your cue from the holiday when you need a midnight snack. Look for dried fruit, cereal, or bananas. High-glycemic-index foods produce sleepiness, so if food must be consumed at night, these are good choices. Other foods that are good choices for sleep contain high amounts of melatonin. These foods include walnuts and tart cherries (dried or juice). Foods high in tryptophan are sleep promoting because tryptophan is the building block of melatonin. Game meat like elk and chickpeas are high in tryptophan content. Finally, foods high in magnesium (almonds) and calcium (milk, kale) can help promote relaxation and sleep. When it comes to sleep promotion, hot chamomile tea or passionflower tea can also be helpful.[56] Sweeten the tea with honey, itself sleep promoting, for an added kick. Avoid proteins that can often promote the synthesis of dopamine, a wake-promoting neurotransmitter.

Given that there are no clear guidelines regarding exactly how much of these foods to eat for improved sleep, I usually recommend eating until the feeling of hunger goes away. Feeling hungry can be a distraction when it's time to fall asleep, so eating just enough to eliminate the gnawing is probably a good guideline.

As with any sleep product, think of these foods as a sleep enhancement. If you are getting to a place where you feel you can no longer sleep without your cup of chamomile tea and bowl full of dried tart cherries, it's time to back up and put these foods in their proper place . . . an option, not a necessity . . . like GPS on a rental car. A little valerian root in your tea from time to time . . . no biggie. A desperate mouthful of over-the-counter valerian pills every night before bed: red flag.

Preparing to Sleep

The room is great, you've decided who's invited, and the glass of wine that you clung to for sleep has exited your life. What should

56 Get Some Zzz's Tea by the Republic of Tea is excellent for sleep because it contains not only chamomile and passionflower, but also 20 milligrams of valerian root, which contains a chemical that has sedating properties similar to the benzodiazepine class of sedatives.

you replace it with? How about a great sleep routine? Think about it. Every kid in the world has a sleep routine:

- Dinner
- Bubble bath
- Dry off and get in pajamas
- Go potty
- Into bed so Daddy can read three books, the last of which is always *Goodnight Moon*
- Quick back scratch
- I love you to the moon. I love you to the sun. I love you to the edge of the galaxy . . . usually ending with I love you to infinity.[57]
- Lights out

Why do kids get to have a set bedtime and a consistent routine, but adults get shut out of the fun? I can't answer that. Everyone can benefit from a bedtime routine, and it can be whatever you want. Routine lets the brain know what's coming. Remember how upset your brain can get when you suddenly fly to Florence, Italy?

For a great bedtime routine, start with exercise in the morning. Exercise is good any time of the day, but if it is done consistently in the morning, particularly in bright light, it can create a positive effect on sleep when it is time to hit the sack. The exercise in the morning, ideally outside in the melatonin-suppressing sun, produces a surge of serotonin that is not only wake and mood enhancing but, if done at the same time every morning, really helps ingrain in your brain, "This is when the day begins." With a consistent wake time, the brain is better able to plan out the next twenty-four hours, including when you will fall asleep. Consistency . . . consistency . . . consistency . . .

With vigorous physical exercise reserved for the morning, try incorporating some relaxing exercises or meditation before bed. Many of my patients "run the list." Their minds are whirring with activity throughout the day, and it's often hard to turn off the to-do

---

57 Which can sometimes be countered with "I love you to infinity and back." This can, in a desperate situation, be trumped with "I love you to infinity times a gajillion."

list at bedtime. So try this: get a notebook and before getting in bed, write down the things on your mind. Practice restricting this writing to an hour-long period in the evening. At any point during this designated hour, you may write things down in your notebook that are on your mind. You do not need to write continuously. Once the hour is over, put the notebook away. You are no longer allowed to think about things you have to do. There's enough on your list already.

This may take some practice! It's a very disciplined practice. Some people find it helpful to not just put the list away, but visualize putting all their intentions and obligations into a box and using a big key to lock it for the night. That kind of visualization can be a useful thing to think about after you lie down to help you fall into sleep.

For those of you with a worry so huge that it cannot be ignored, go ahead and write it down. It's not that big of a deal. The worst-case scenario is that you'll awaken, turn on a dim light, write it down, and struggle to fall back to sleep. Oh well, at least you won't miss your tax filing deadline tomorrow! A trick for these situations is to keep an unusual (and unbreakable) object on your nightstand. Perhaps a wooden carving of St. Elijah, the patron saint of sleep.[58] If you awaken during the night and think of something important, grab the object and toss it onto the floor. When you awaken the next morning and see St. Elijah there, you'll think, "Why is he there? Oh right, I need to make sure I buy my Powerball ticket today" or whatever big thing was on your mind.

 PRODUCT SUGGESTION

WANT TO TURN THE PROCESS of quieting your mind into a futuristic video game on your iPhone? Muse (www.choosemuse. com) is a small biofeedback device that wirelessly connects to your phone. Muse detects your brain waves and can convert them into the sound of the ocean. By using the device, you can "hear" the activity level of your brain and practice calming it

58 Hey, it can't hurt!

down. The more you calm your brain, the quieter the ocean becomes. By practicing this skill, you'll be able to expertly shut
your brain off as soon as you get into bed.

One useful component to the bedtime routine is a hot shower,
or preferably a hot bath. While cooler environmental tempera-
tures typically result in a higher-quality sleep, vigorously heating
the body via bathing before sleep has been shown to improve sleep
quality, most likely because of the cooling and body heat release
that follows. Therefore, a hot bath about an hour before you hit
your cool, comfy bed can be very helpful for difficult sleepers. This
is consistent with recent studies that seem to point to a link between sleep and temperature that is much stronger than previously thought. These studies show that relatively small changes in
environmental temperature leading to reduced body temperature
often lead to improved sleep.

Funny story. When my son was little, he skinned his knee riding
a scooter. While the wound was not serious, it did need to be dealt
with, and he was aggressively disinterested in its treatment. Conversations about gangrene and lost limbs had no effect, so I decided a sneak attack was the best option. When bath time came
around (in his perfect bedtime routine), I suggested we take a bath
together in the big bathtub and play Playmobil pirates. He was
thrilled, and before I knew it, we were in the tub divvying up the
plastic buccaneers. While I was working on a plan to clean the
bloody, crusty knee wound, my son was giving me all of the crappy
figures and broken swords and keeping all of the really cool pirates and boats for himself. As the war for the treasure chest began, I would attack his fleet with a figure in my left hand while my
right hand would splash water and soap on his knee. While I lost
the pirate battle miserably, I clearly won the knee war. But I had to
keep doing it every night, since he never did get more agreeable
about washing his wound after a busy day of play.

That entire week, I kept falling asleep early, so early that my wife
even commented, "What's going on with you?" I had no idea. It
took me several days to realize my sleepiness was probably related
to the early hot bath I was taking!

With all this in mind, an adult's sleep routine might look something like this:

- Exercise in the morning, preferably in bright light.
- Consistent breakfast timing with the meal heavy on wake-promoting protein.
- Consistent lunch timing.
- Finish dinner at least three hours before bedtime. If you have to snack after that, have a handful of nuts or a few pieces of dried fruit. Not too much.
- Reduce your environmental lighting around the time the sun is going down. Turn off lights or use dimmer switches.
- Spend an hour after dinner jotting things down on your to-do list. Put the list away after sixty minutes.
- Brush your teeth.
- Take a warm bath.
- Do some light exercise or meditation. Breathe deeply.
- Read a print book until you feel sleepy.
- Turn out the light, and snuggle into a cool bedroom environment.

Final tip. You've heard the adage "If it takes you more than twenty minutes to fall asleep, get up and go do some quiet activity until you feel more sleepy." I have no big issue with this tip, but I do have a few suggestions:

1. Forget the twenty-minute rule. It's just an arbitrary number. It could have been seventeen minutes. What I don't like about this tip is that it is putting the sleeper, who is already struggling, under a new pressure. "I better fall asleep in the next twenty minutes or I . . ." Who cares if you fall asleep in twenty minutes? If you do, bully for you! If you don't, you are still okay. Instead of setting an arbitrary time like this, just try to pay attention to your body. If you've been in bed for a while and you don't feel like sleep is going to happen anytime soon, you can get up.
2. But you don't have to get out of bed if you don't want

to. If you are still "trying" to fall asleep even though I told you not to, and this is stressing you out, it's okay to get out of bed. If you are not too bothered by the situation, I recommend simply lying there and resting. Plan out your dream vacation. Plan a surprise date with your partner or a thoughtful gift for a coworker. It's important to remember that resting even without sleeping is good for you too. You're not wasting your time if you are lying in bed and not sleeping.

3. If the twenty-minute thing is happening frequently, listen to your body. You are going to bed too early. It's time to go to bed a little later.

Entire books have been devoted to the contents of this chapter. I don't want to belabor sleep hygiene. Sleep hygiene is like Donder the reindeer: an integral part to the overall Santa sled operation, but definitely not the whole story or even the most important piece of the puzzle.

....

CHAPTER 8 REVIEW

1. Make your room dark. Crazy dark.
2. Spend so much money on your new bedding and room furnishings that you can't afford your nicotine and caffeine habits anymore.
3. Buy your spouse a special gift, then kick him out of your bed until his own sleep problems are fixed.
4. Develop a sleep routine. Feel free to incorporate *Goodnight Moon*. It works on adults too.

I sincerely hope that your sleep problem was fixed with simple caffeine reduction and light removal from your bedroom, but it's okay if it wasn't. Sleep issues are usually much more stubborn and deeply rooted. Read on and start understanding how to deal with your insomnia on a higher level.

9

INSOMNIA
I Haven't Slept in Years, Yet I'm Strangely Still Alive

HOPE TO GOD YOU DIDN'T just skip all of my hard work writing this book and jump right to the chapter about insomnia. If you did, I highly suggest you go back and read everything. It's important. Come on, it's not that long, and given how little you tell everyone you sleep, let's face it: you've got the time. Don't worry. I'll wait for you right here.

There is one important thing we must recognize when it comes to the diagnosis of insomnia. For the most part, you, the patient, are completely in charge. Mull that over. The diagnosis is made by the patient, not the doctor. In other words, the patient gets to decide if she has the condition . . . not the doctor after performing an exam. Not the results of a blood test or MRI. The patient.

"Good morning, Doctor. I have terrible insomnia and I don't sleep."

Quick: Name a condition that a patient gets 100 percent control over the diagnosis.[59] Imagine what would happen if I walked into my doctor's office and said, "My chest hurts. I'm having a heart at-

---

59 Sorry. Gluten sensitivity doesn't count.

tack. Please give me a stent" or maybe "My abdomen feels bloated. I'm clearly pregnant."

This lack of objective evaluation often gets the treatment of insomnia off on the wrong foot. Imagine a patient who feels he or she does not sleep that much, but actually does. (Remember sleep state misperception in Chapter 6?) If the patient is in control of the diagnosis and language used to define it ("I can't sleep"), how effective will the doctor's sleeping pills be if the patient is already sleeping?

So if insomnia is not "not sleeping," what is it? It's simple. It's not liking your sleep. You are allowed to not like your sleep yet still be sleeping. You can dislike your job, yet still show up for work every day. I think it is okay for a doctor to help a patient understand that his insomnia is not stemming from an absence of sleep. The doctor is simply reframing and redefining the problem. This redefinition should *never* lead to dismissing the patient or the problem. This is so important, I'm going to say it again in a slightly different way:

Knowing that a patient who says he can't sleep does in fact sleep is *not* the same thing as the patient not having a problem or not needing treatment.

The concept of sleep state misperception (or paradoxical insomnia) is not an excuse for not treating a patient. It merely is a tool for providing a way to better define and treat a patient's sleep problem. This patient is seeking help or buying a book for a reason. In the case of the insomnia, the patient may need help figuring out exactly why he doesn't feel his best.

Let's consider the term *insomnia*. How do I define it? Before I give you my definition, let me share with you how most people define insomnia:

"It's that thing when someone can't sleep."

Wrong! We've already established that everyone sleeps, sometimes. A better definition involves two key elements:

1. An individual who is dissatisfied with the quality of his sleep on a regular basis, say two to three times a week or more for three months. These lines in the sand are arbitrary. If you have difficulty sleeping once a month, and it really bothers you, then in my opinion, you have insomnia and we're here to help. I would remind you that a difficult night of sleep from time to time is okay. It's part of the human condition, so to speak. We have really bad breakups with girlfriends. Our pet rabbits die unexpectedly. Fantasy football quarterbacks underperform on *Monday Night Football*. To put it bluntly: Shit happens. If that difficulty is happening more than what you like or can tolerate, you are halfway to having insomnia.

2. An individual who cares, and cares a lot. Having difficulty with your sleep from time to time is not insomnia. To truly have insomnia, that difficulty sleeping has to bother you, annoy you, really get under your skin. It was shown in an interesting 2012 study that insomnia patients remember their bad nights of sleep more than their good nights. This selective memory is often on display when I ask a patient how her sleep has been since her last visit two months earlier. "Terrible" is the reply, but after looking at her sleep diaries, I see there are often more good nights than bad. For some insomnia patients, it is almost as if the good nights had never happened. For healthy sleepers, the opposite is true. They pay little to no attention to the night or two of bad sleep they have.

I think that misperception can happen to some people because they find not being able to fall asleep intolerably frustrating. They might even describe it as "terrifying." Not falling asleep makes them so anxious, they can't fall asleep and are left feeling so "helpless" (even though their own reaction to not sleeping is what's causing their problem) that they are truly scared. It happens a lot to people. I get it. The strength of that fear is part of the reason it's so important for people to realize that they are sleeping. But I also think it's part of the reason these patients don't notice the nights they sleep well. Their focus is on a kitten in the room playing with a ball of yarn, yet they perceive a tiger.

Pulling this all together, we can create a definition of insomnia that is simple and complete. Insomnia is not when an individual can't sleep. The true definition of insomnia consists of two components:

1. A person is not sleeping *when she wants to sleep.*
2. The person cares, and usually cares a lot, about not sleeping, whether or not she wants to admit it.

Let's first examine number 1. There are many ways people cannot sleep when they want to sleep. For example, sleep-onset insomnia is when an individual struggles to fall asleep. Most people consider a thirty-minute struggle or longer to meet the insomnia criteria. I think any amount of time, if it is frustrating, fits the bill.

Others struggle to maintain their sleep. In sleep-maintenance insomnia, sleep may initially come quickly, but prolonged awakenings punctuate the night. Classically, the sleep-onset insomnia patients were thought to be anxious, whereas the individuals who struggled to maintain their sleep or who awoke too early were thought to be depressed. Most sleep experts do not subscribe to this way of thinking anymore. In reality, a good way to think about this is as follows: Anyone whose sleep is inefficient—meaning the time he sleeps divided by the time he is in bed is a low number, say less than 75 or 80 percent—has insomnia.

Let's look now at component 2 and the way individuals react to difficulty sleeping. When I go to bed at night, I virtually always fall asleep immediately . . . not always, but virtually always. On the nights when I get into bed, turn out the light, and don't sleep immediately, I genuinely don't care. I don't fear the situation. I don't anticipate it having any real consequence in my life. I'm doubtful of it happening two nights in a row. I sometimes challenge myself to see if I can lie quietly in the bed all night without falling asleep. I think about fun things to plan for the weekend, my family, having Giada make me a glorious and rich Italian dinner, and other important things. I've never even come close to making it all night, but even if I did, or even if you do, remember that there is benefit to resting. In 2005, neuroscientist Gilberte Hofer-Tinguely showed resting without sleeping improved cognitive performance. Resting is not wasted time; in fact, a 2009 study revealed that for some cognitive tasks, the benefits of resting are indistinguishable from those of sleep. So don't worry too much about getting into bed and not sleeping immediately or having a prolonged awakening during the night.

Before we continue, I want to take this moment to remind you that while I am a genuine board-certified neurologist and double-board-certified sleep specialist, I am not a conventional sleep doctor. I've been in the sleep business long enough to understand that sometimes conventional thinking is not always the best approach. I think most "good" doctors think this way. No two patients are alike, so why try to force them into rigid boxes?

My ideas about insomnia are not always conventional.[60] I do not find the current way sleep medicine chooses to organize and deal with insomnia to be very helpful, although it is getting better. We used to divide insomnia up into a million subcategories: insomnia with inadequate sleep hygiene, insomnia with sleep state misperception, insomnia for people with chronic medical illnesses. The list went on and on. I have found these divisions useless in treating patients because patients often display elements of many insomnia subtypes. Currently, the American Academy of Sleep Medicine uses a classification system that is more useful: short-term insomnia, chronic insomnia, and other, with *other* being people who haven't decided if they are short-term or chronic yet.

While this is a glorious step in the right direction, I think it can be even better. In this book, we will do away with "other" and focus on either short-term insomnia or chronic insomnia. However, to me, the fuzzy time classifications that distinguish acute, or short-term, insomnia from chronic insomnia are unhelpful too. Virtually everyone has or will experience acute insomnia from time to time. It's like itching. Sure, itching is an unpleasant experience, but we all get a weird, isolated itch from time to time. We scratch it and move on. If that itch persists and keeps coming back, then it becomes chronic

60 Prepare yourself. Seriously. You may not like what I have to say in a few pages.

and the hunt begins for its cause. Same thing with insomnia. So for this book, we are going to classify insomnia differently. I've come up with the labels "simple insomnia" and "hard insomnia."

Simple Insomnia

Humans worry and stress about things, even trivial stuff like melting glaciers and water shortages. These concerns, and others, can occasionally lead to poor sleep. I consider a night or two of sleep problems here and there to be normal. Who cares? If that's the case, why even have a diagnosis of simple insomnia?

To me, the biggest reason for creating the category of simple insomnia is to reinforce the idea that it's just that. Simple. Harmless even. Most likely the cause is staring you in the face. I put this section in the book so people who develop simple insomnia can learn to recognize it early and nip it in the bud before it becomes hard insomnia and more difficult to treat. *Simple* also implies optimism. You can easily fix this.

To me, the key to simple insomnia is doing a thorough inventory of its causes to see what factors might be contributing to your problem, and then making them go away. There are many causes for simple insomnia. Scores of articles out there painstakingly go through every possible cause for not sleeping. I would be willing to bet anything that for many insomnia sufferers, this content is either not relevant to their situation and/or items they have already considered and dealt with. Now that you are this far into the book, I know that you, Reader, know too much alcohol is harmful to your sleep. You've tried the melatonin. You have a consistent sleep routine. And yet, your problems with sleep persist. In other words, you followed all the advice in Chapter 8. So why are you still having trouble sleeping?

Anxiety

Time and time again anxiety usually emerges as the number one cause of insomnia. Don't believe me? Go find an insomnia blog.

Now go find a blog for people struggling with some rare and deadly tropical disease. Which group seems more anxious about their condition?

Insomnia expert Charles Morin theorizes that people with slight anxiety tendencies are more likely to struggle with insomnia. They are *predisposed*, to use his word. Being type A has its perks, but type As also make really terrible sleepers a lot of the time. Their little hamster-wheel minds just won't shut off. At one such insomnia blog, the people leaving comments used the word *mind* more than fifteen times:

> "keep your mind alert"
> "quiet your mind"
> "root of it all is a wandering mind"
> "truly relax my mind"
> "mindfulness isn't easy"
> "mind works overtime"
> "I need my mind engaged"
> "shut my mind off"
> "mindfulness stress reduction clinic"
> "mind over mood"
> "anxiety-prone state of mind"
> "mindfulness-based stress reduction"
> "practice mindfulness"
> "clearly mindfulness is the core component"
> "I can't sleep because of my racing mind"

What is a racing mind? Try Googling "racing mind" or "racing thoughts" and see what you come up with. Words like *bipolar disorder, mania, OCD*, and *anxiety* are everywhere. Look, I'm not saying because you have problems with your sleep, you are bipolar, but just start opening yourself up to the possibility that there could be some anxiety issues lurking around in your bedroom.

Sleep is a skill to some degree. We all eat, but some among us can eat forty-three hot dogs in ten minutes. These individuals have taken an act that we all engage in and, through training, have made it into a superior, nauseating skill. Like anything else we do, we can approach our sleep in a similar way. We can learn to be good at it.

Our minds are often in the way when it comes to our sleep. The more we concern ourselves with sleep, the harder it is to initiate it. When Tiger Woods was young, his father, Earl, would try to spook him before shots or create situations with artificially inflated pressure to acclimate Tiger to stressful circumstances and help him cultivate the skill of blocking out distractions and focusing on the task at hand. I'll go on record now by saying that the skill involved in falling asleep when you're sleepy is much easier than sinking a forty-foot putt to win a tournament with millions on the line. As the champion bull rider Donnie Gay once said when talking about riding a 2,000-pound bull, "There are different degrees of pressure." But, honestly, for people who suffer from insomnia, the pressure to fall asleep can feel just as daunting as making that putt or riding that bull.

Once an individual has suffered with a sleep problem for more than three to six months, changes in the psyche of that individual start to happen. Going to bed, a relatively benign activity, starts to become a very negative task. Hours before one goes to bed, dread of sleep starts to set in. People start to wonder if they have enough sleeping pills. They resent their significant other's ability to fall asleep quickly. Frustration rapidly sets in as sleepers toss and turn in a futile effort to fall asleep. For some sufferers of chronic insomnia, the event that led to the insomnia in the first place becomes irrelevant. Many people go through a period of poor sleep during a divorce. To think that divorce is what is ruining your sleep ten years later is ridiculous. I have patients tell me all the time that their years of insomnia were caused by a job loss in the distant past. But if it truly is in the distant past, that's simply not true.

To understand why, it is helpful to understand how insomnia comes about. Patients often have events like divorce or job loss that cause a sudden increase in anxiety and result in a sudden change in their sleep quality. Some people rebound from these difficult life events and continue sleeping fine after a brief hiccup of poor sleep. For others, the poor sleep continues but usually not because of continued anxiety about the initial event, but rather now because of concern about their sleep. In other words they have problems sleeping because they worry about their problems sleeping.

People who suffer from insomnia often put mental pressure on themselves to get to sleep at night. Many people worry that if they "don't get to sleep sooner" their productivity will suffer at work or they will feel exceedingly bad during the day. Their anxiety about the consequences of not sleeping ratchets up to fear, and before long, they are in such a state of vigilance that they are as far from sleep as it's possible to be.

But bad sleep is much more dangerous in your mind than it is in real life. I've spent many nights staying up working on research projects, doing my taxes, or performing more mindless tasks. I've turned around and gotten up to exercise before the sun comes up. Do I feel amazing the following day? God, no. Can I make it through the day and be productive? Sure. Just don't cut in front of me in line that day; I'm prone to bite your face off. Insomnia patients often use words like *dysfunctional* to describe what happens if they don't get a good night's sleep.[61] Just because you got little to no sleep the night before doesn't mean you can't handle the next day. I'm not saying that your day will be chocolates and roses. I just don't think you'll end up being truly dysfunctional.

Medical Issues

Illness and medications used to treat illnesses can also be responsible for insomnia. These illnesses include physical disturbances, often featuring pain, and psychological problems like acute anxiety and bipolar mood disorder. The medications used to treat individuals for conditions like these can be an independent cause of insomnia. Common among these are steroids, antidepressants, and allergy medications.

With medical issues, it is important to understand the difference between primary insomnia and secondary insomnia. Primary insomnia involves sleep disturbances that do not have an obvious cause. Secondary insomnia is a sleep disruption with a definable cause. For example, imagine a patient with significant leg pain. This person may struggle to fall asleep at night because of the searing pain in his ankle and big toe. Is this really a sleep problem?

---

61 So much so that I refer to the word *function* as the "F word" in insomnia clinics.

To me, if an individual came to my clinic with a bear trap clamped onto his ankle, I would consider this less of a sleep problem and more of a "bear-trap-on-your-ankle" problem.

Cognitive Behavioral Therapy

Stress and anxiety are everywhere. Some of it you can control (leave an unhealthy relationship, give up rooting for the Cleveland Browns) and some of it you cannot. Identifying and managing stress, the very definition of cognitive behavioral therapy (CBT), is an important step toward improving your insomnia.

For some, the insomnia and anxiety are so tangled up and chronic that a more focused approach needs to be taken. This is the main reason a large meta-analysis published in the *Annals of Internal Medicine* in 2015 showed cognitive behavioral therapy to be a very effective treatment for insomnia because it goes after the bad beliefs, anxiety, and poor habits central to the condition. In my opinion, no other therapy comes close.

There are books and books out there that focus only on cognitive behavioral therapy. I'm not going to do the technique a disservice by trying to reproduce it in a few pages here, but I think it's worth hitting the highlights, many of which you probably already know.

What Is Cognitive Behavioral Therapy?

Basically CBT involves an approach to insomnia, or another psychological disorder,[62] that involves changing the way you sleep by diving deeper into the mechanisms and behaviors that might be leading to the insomnia or worsening it. Cognitive behavioral therapy can be geared to many things: for fear of flying, for test-taking anxiety, for any debilitating but irrational fear you might harbor. When CBT is employed specifically for insomnia, it is sometimes designated CBT-I.

---

62 Yes, I said "psychological." That's not the same as saying you are crazy or that you are making this up. I'm using the word *psychological* in its purest sense; our mind is creating a problem.

There are several components or techniques that fall under the CBT-I umbrella, each one designed to help with the ability to initiate sleep.

- **Good Sleep Education:** This component is not always pulled out in discussions about CBT-I, but in my opinion, it is essential. Patients need to understand the science of sleep. They need to understand what is real and what is theoretically impossible. If a patient tells me that he can stand in front of the sun and photosynthesize food like a rhododendron, we are not going to get too far until we come to an understanding that as far as science can tell, this is impossible. In terms of CBT-I, the aim of this entire book is to educate you about sleep in general so you can better understand your own sleep patterns. Welcome to CBT-I. You've already started and didn't know it!

- **Good Sleep Hygiene:** You know all about this already, and you have your cozy bedroom, fluffy pillows, and comfy pajamas.

- **Stimulus Control:** This is a very formal way of saying the bed is for sleep and nothing else. This includes, but is not limited to, studying in bed, working in bed, doing your taxes in bed, even doing the *New York Times* crossword puzzle in bed. You know that already! Wow, what a waste of time this section is becoming. You know too much. In addition, stimulus control mandates that you make your bedroom as inviting and sleep conducive as possible and that you go to bed only when sleepy.

- **Sleep Restriction:** Basically, figure out how much sleep you need and give yourself that time in bed. If you seem to always take a long time to fall asleep, stop spending so much time in bed. I'll be honest. I think *sleep restriction* is a bad term. It really should be "time-in-bed-not-sleeping restriction" or maybe "thumb-twiddling restriction." Whatever . . . nobody ever asks me my opinion before these things are named. Sleep restriction is so important and misunderstood that I gave it its own section.

▪ **Relaxation Training:** Remember my definition of insomnia, specifically part 2? You have to care about insomnia's effects for it to rule your life. It's easy for me to tell patients to relax, but often difficult for people to do so. These techniques are utilized to help individuals learn how to chill at night. Start with your toes. Stretch them, and feel them relax. Now your calves. Have you done this before? You gradually work your way up your whole body, part by part, relaxing and breathing deeply. This technique is great because it not only helps patients relax but it gives their brain a plan for going to sleep and not nodding off immediately. In other words, the old plan of "get into bed and feel miserable as I *try* to sleep" is replaced with "get in bed to rest and do my relaxation drills." Remember: Never *try* to sleep.

▪ **Cognitive Therapy:** This is the key ingredient. If CBT-I is a chowder, cognitive therapy is the clams. This aspect of the therapy is targeted at eliminating or altering a patient's irrational or unhelpful beliefs about sleep. "When I don't sleep, I can't function."[63] Cognitive therapy would say, "You woke up, you taught your third-grade class, you went grocery shopping afterward, and while you didn't make it to the gym, your day was not *dysfunctional*." This cognitive therapy also works toward helping patients worry less. Insomnia without worry is like Gollum without his precious ring: weak, pathetic, and powerless. (Cognitive restructuring is discussed more in the next chapter.)

Parting Thoughts Before Moving on to Hard Insomnia

I hope reading this book has helped you both understand your sleep more fully and figure out solutions to your problems. While not traditionally so, this entire book was written with an eye to-

---

63 There's that *function* word again.

ward CBT-I. Despite my careful planning, unfortunately some people reading this book are not going to find quick solutions to their insomnia. That's simply a fact. Doctors are human. Medical resources are limited, and some people, no matter what happens, feel like they cannot sleep from time to time. One incredibly powerful tool in your fight against sleep disturbances is acceptance. Accept your sleep for what it is, optimize what you can, and move on with your life.

I have seen thousands of patients with sleep issues and insomnia. In my experience, the disturbance is as debilitating as an individual chooses to make it. Let me explain what I mean.

Visit any university teaching hospital at night. In fact, for a real thrill, go back twenty years, before work hour restrictions were put into place. Talk to a doctor who took calls during that time. I remember my residency being one of virtually no sleep when we were on call. That was the norm. Residents were going with limited or no sleep every other night or every third night for months if not years. Take a look at the level of functioning of these people. It was really high. These individuals were operating, doing spinal taps, sticking lines into patient's necks, that kind of thing. Highly functional? Absolutely. Sleepy? God yes. But the bottom line was this:

Despite extreme levels of sleep deprivation and sleepiness, these individuals functioned surprisingly well.

Why is it then that patients with insomnia, who often demonstrate virtually no discernible sleepiness, are so burdened by the disability of their sleep disturbance? Perhaps because it is a choice. If this book helps you improve your sleep, then I have been successful.

If it doesn't, kind Reader, I sincerely hope you make the choice that while you are working on your sleep's improvement, the sleep difficulties will not ruin your life. Make the choice that you are going to feel great tomorrow regardless of your sleep tonight. And if tonight's sleep is not amazing, resolve that tomorrow's will be.

Don't make your sleep disturbances a defining characteristic in your life. The hour or two it takes you to fall asleep is not that big of a deal. Believe this. Free yourself. You are in your comfortable bed, away from the stresses of the day, stretched out and relaxed. Is this a situation to fear and get upset about? Don't let this small issue lead you down the dark path to hard insomnia.

 ## THE INSOMNIA EQUATION

I'VE COME UP WITH AN algorithm that will predict how long it will take to solve your insomnia problem.

$$\frac{1+ (\text{Insomnia Years}) + (\text{Sleeping Pills})}{(\text{Hours You Sleep/Night}) \times (\text{Epworth Score})} = \frac{\text{Months It Will Take}}{\text{to Improve}}$$

Insomnia Years: How many years you have had insomnia

Sleeping Pills: How many sleeping pill brands you have tried

Hours You Sleep/Night: How many hours you average sleeping every night

Epworth Score: Fill in your Epworth Score from Chapter 3.

Note: If your score produces "error" or "cannot divide by zero," you have either not read this book in its entirety because you still think you sleep 0 hours per night, or despite your inability to sleep, you feel absolutely no sleepiness. In that case, I'm way sleepier than you . . . You should be helping me, not me helping you.

CHAPTER 9 REVIEW

1. Anxiety and stress are key components of insomnia. Accept that they are playing a role in your sleep disturbances. Work to minimize them.
2. It is essential that you take an honest look at all of the factors contributing to your poor sleep and develop a plan for improving them. Enlist the help of others. Be open to their perceptions.

3. Develop a CBT-I plan, and if you cannot do it yourself, know there are certified CBT-I providers sitting in little offices everywhere just waiting to help.

Did you calculate your Chapter 9 exercise score? Prepare yourself. Fixing this problem takes time, and the longer it has had to sink its roots into your psyche, the longer it will take to extract. Settle in . . . I named it hard insomnia and not "impossible" insomnia for a reason.

10

HARD INSOMNIA
Please Don't Hate Me
When You Read This

WAS WATCHING A MEDICAL SHOW recently and a woman came on who said she had not slept for twenty years. No sleep since 1995 is a long damn time. The sleep expert on the show, resplendent in her starched white lab coat, smiled knowingly as she delivered her solution to this haggard woman. The woman was in luck as this expert was kind enough to bestow upon her two beautiful glowing pearls of sleep wisdom so precious and powerful, this woman would surely sleep the sleep of angels that night. And like a faith healer telling a member of her congregation to rise and never use that wheelchair again, she pronounced the following:

1. Find the most boring instruction manual you can find and read it while in bed.
2. Turn your body around 180 degrees in the bed so that your head is at the foot of the bed and your feet are at the head.

A couple of things.

First, if this advice is all it takes to solve that poor woman's sleep problems, I'll switch careers immediately. Judging by the look on the woman's face, her sleep problems were probably headed for year twenty-one.

Who are we kidding here? These are simple insomnia tips being lobbed at a woman with hard insomnia. We are fighting the Incredible Hulk with a goddamn slingshot.

Second, how are we going to solve this person's difficulty sleeping when we know so little about her? I wrote this book with the intention of it being something I would give my patients—an extension of our clinical visit. While a tip like "put your head at the foot of the bed" is perfectly harmless, I think most doctors would agree that advice like that would have little chance of working on a patient like this—a patient with hard insomnia. She could paint her bedroom a cool blue, a color thought to be relaxing. But the only way a bucket of paint is going to help this woman sleep is perhaps if she inhales the fumes for a while and passes out on her bed. Tips like this are fine, but within the population of individuals who cannot sleep, I think such quick fixes are frankly insulting and set patients up for more failure, something they desperately do not need.

Hard insomnia definitely deserves its own chapter and probably its own book. Hard insomnia is a soulless beast that sucks all the hope and happiness out of an individual's life. Okay, fine, it's not *that* bad, but if you talk to an individual who struggles with hard insomnia, you soon learn it can be pretty rough.

Insomnia in general is an interesting thing. For the most part, it's a symptom, not a condition. In other words, there is no such thing as sore throat syndrome. If your throat is sore, it is because you have strep throat or a viral pharyngitis, or you've been screaming all night at a Justin Bieber concert. This does not stop patients from talking about insomnia like it is an inherited trait, like missing adult teeth. I still have a baby molar. Bring this book to a book signing and I'll be happy to show it to you. Want a real treat? Ask my mom about hers. I think she's got three. I guess I get my retained primary teeth and absent secondary teeth from her.

Insomnia does not work that way. There is no known gene for insomnia, but there are most likely genetic factors at play in the development of an individual's difficulty with sleep. In other words, there is no gene responsible for dunking a basketball, but a gene influencing height might be strongly correlated to dunking ability. Does it mean short people can't dunk? Not necessarily. Does it mean tall people are always able to dunk? Not at all. These nuances were a bit lost on the media when Dutch sleep researcher Eus Van Someren's research about an "insomnia gene" hit the Internet. Prepare yourself for what is to come.

Extra! Extra! Read all about the study that basically said, in the text, that these individuals sleep—in fact, they can sleep quite a bit. Their sleep is often quite fragmented, though, which once again is a very different thing from "not sleeping."

I'm totally open to the idea that some individuals possess genetic programming that might influence their chance of success with sleeping. I'm also, however, open to the idea that programming can happen throughout an individual's life that has nothing to do with genetics. Imagine a kid waking up to his mother complaining every morning that she couldn't sleep and that she felt miserable. That kid might have the thought as he is eating his Lucky Charms that whatever is going on with his mom might be going on with him on the rare night he sleeps poorly. So is there a true insomnia gene? I'm going to say it's doubtful in the way we think about genes causing eye color and the ability to curl your tongue. Is it probable that some people are more "insomnia resistant" based on their genes? Absolutely.

Because of these factors, we often talk about insomnia as being either primary or secondary. We covered this in the last chapter, remember? Secondary insomnia is when your insomnia is the result of some other condition or factor. Chronic pain is a common one. Let's say you have a lightning-like pain that shoots down the back of your leg from buttock to big toe. That sciatica pain burns like hellfire during the night, which makes it really hard for you to fall asleep. That really isn't a sleep problem you have; it's a pain problem that causes the secondary problem of difficulty sleeping.

The cause of hard insomnia, however, can be difficult to pin down or, in some cases, seemingly impossible. We often refer to this as primary insomnia: insomnia with no clear cause. This is where insomnia can become a very dark and bleak condition.

I'm going to warn you, kind Reader, that many of you are not going to like what I have to say. In fact, if you grabbed this book off of the shelf and jumped right to this section, you may be about to get very upset. Try to keep an open mind as you read.

I deal with chronic insomnia on a daily basis and have done so since about the time Michael Phelps began hoarding Olympic gold medals. I see insomnia all of the time, most days of my life. Insomnia patients are frustrated, desperate, and so unbelievably sick of having problems sleeping that they are at the end of their rope. I don't think I'm going out on a limb when I say that these individuals can be described as traumatized. Let me be clear. I'm not saying that they are traumatized because they aren't sleeping. What I'm saying is that the presence of hard insomnia for years and years is itself traumatizing.

Most patients with chronic insomnia have been "bad sleepers" for years. They have tried many different kinds of medications, usually with little to no sustained success. In fact many are taking drugs that they freely admit do little to nothing to ease their difficulties. Weigh that fact for a minute. They are taking drugs that don't work. What on earth would drive such behavior? I've never seen a blind person wear glasses that don't help her see.

Hard insomnia sufferers have seen doctors on top of doctors on top of doctors. They have seen hypnotists, counselors, acupuncturists, massage therapists, and biofeedback specialists. They blog. Dear God, do they blog.

I have attended more than my fair share of sleep conferences. I have lectured and presented research at these meetings, but more often than not, I am listening to the work and research of people smarter than I am. I'm going to let you in on a secret. The way in which doctors talk about chronic insomnia patients during these closed medical sessions is *not* the way they talk to chronic insomnia patients in person or in print. I've never been to an orthopedic conference, but I'll bet the way bone doctors talk about broken legs to each other is pretty much how they talk to you.

Insomnia Distress

I have three kids: one daughter and two sons. My daughter is a high school senior now and most likely is heading off to college somewhere. It has been an interesting experience watching and guiding her through school.

It is well established that the way all students, particularly girls, view their abilities in math and science greatly impacts their performance in these subjects. This identity as a "good math" student or a "bad math" student can be formed at a fairly early age. Once a bad-math-student identity is formed, despite an overall ability in math that is high, this student will fail and ultimately avoid careers that feature math and science. The course of math avoidance does not make sense when you look at the grades of these students. Often girls perform just as well as their male peers in the classroom. In other words, their performance does not reflect their abilities or even their testing outcomes.

There is a similar phenomenon seen with the insomnia patient. Kenneth Lichstein, a sleep researcher, calls it the "insomnia identity," and I think this is a brilliant term. The insomnia identity centers around the idea that the insomnia patient believes himself to be a bad sleeper or someone who cannot sleep, often despite evidence to the contrary. This is the main target of CBT-I.

I remember working at Camp Holiday Trails as a counselor back in the mid-1990s. The doctors at the camp made it very clear we were not to refer to campers with diabetes as "the diabetics" or the kids with blood clotting disorders as "the hemophiliacs."

Why not? It seemed logical to me.

"Because these are amazing kids who are so much more than this medical problem. It does not define them, and we don't want it to start doing that. So instead of the diabetic kid, he is the kid with diabetes." It's a subtle, but all-important difference.

Many patients with insomnia are "insomniacs" rather than simply "people who have trouble sleeping." On the rare occasions I have a bad night of sleep, I do not for a minute consider myself an insomniac. Why would I? Am I forgetting about all of the other wonderful nights of sleep, the lazy naps taken on vacation, or the embarrassingly deep drool-on-yourself slumber I get on the many flights I take around the world every year? Does this sleep not count?

Of course it does, but for the insomnia-identity patient, the self-image of being a bad sleeper is not affected by little things like reality and facts. In fact it is well documented that insomnia patients will often ignore their good nights of sleep—the ones during which they slept seven hours—and report only the difficult nights.

When it comes to hard insomnia patients, we must work in a reality-based world with blue skies and green grass. Going back to the math student, let's look at her grades. Wow, all As including As on two tests and six quizzes and one C on a homework assignment! She is a fantastic math student! It is very important for her to know this, as it could color her confidence and self-identity forever. Yes, she got a C, but she understands the mistakes she made on the assignment, so it's not a big deal. Certainly nothing to focus too much attention on.

Insomnia patients in many cases have made up their minds that they are bad sleepers, and with this identity usually comes the disability.

With this in mind, we can construct a chart for viewing all sleep patients:

FIND YOUR SLEEP IDENTITY

		SLEEP QUALITY	
		EXCELLENT SLEEP QUALITY	POOR SLEEP QUALITY
SLEEP IDENTITY	Positive Sleep Identity/ Low Distress	Engaged and normal sleepers (Good sleepers and they know it)	Disengaged, poor sleepers (Poor sleepers, but think they sleep well)
	Negative Sleep Identity/ High Distress	Disengaged good sleepers (Good sleepers, but feel their sleep is poor)	Engaged, but poor sleepers (Poor sleepers, but aware of the problem)

In looking at this chart, you can see there are individuals who sleep poorly and identify themselves as poor sleepers. Likewise there are the annoying people who are awesome sleepers, know it, and usually love to tell you all about it.[64]

Now pay attention to the gray boxes. These are the individuals Lichstein and others have labeled disengaged sleepers. Their sleep complaint is disengaged from their sleep reality. These are people

---

64 If you are one of these people, why are you spending your money on my book? You should treat yourself to a massage or something!

like my sleep apnea patients who are horrible sleepers. They choke, cough, kick, groan, and wheeze all through the night, keeping their partner up, yet have no idea why they are being sent to my clinic, even after falling asleep in my waiting room. They think they are excellent sleepers.

There is another group of disengaged sleepers. They are the sleepers who sleep well through the night. Their sleep is ample in amount and of seeming excellent quality. Often we see these patients in conjunction with their sleep study that confirms pretty awesome sleep. Despite this seeming excellent sleep, these individuals are beside themselves because their sleep quality is, in their minds, so poor.

Studies have examined these different sleepers. One study in particular looked at good sleepers compared to poor sleepers who exhibited low distress ("disengaged") and poor sleepers who exhibited high distress. These groups were compared based on:

1. Their sleep quality
2. Their self-reported fatigue, their sleepiness (remember the difference!), and their cognitive abilities[65]

In two separate groups studied in 2000 (136 college-age subjects and 194 older adults), the high-distress and low-distress poor sleepers had virtually the same sleep quality, which was far worse than the good sleepers'. However, when it came to their functionality, the high-distress poor sleepers had more depression, sleepiness/fatigue, and cognitive impairment than the low-distress poor sleepers. Moreover, the low-distress poor sleepers seemed to function at a comparative level to the good sleepers based on these measures. In other words, to feel great, you don't have to sleep well. You just have to believe you do!

Unfortunately, the opposite can be the case as well. To feel poorly, you don't have to sleep poorly (or sleep too little); you just have to believe you do. This too was seen in studies. Good sleepers with high distress functioned more poorly than good sleepers with low dis-

---

65 The importance of cognitive impairment in insomnia cannot be overstated because most of my insomnia patients report being mentally "dysfunctional" at work, often with nothing really to show for it.

tress. No surprise there. What was surprising was the high-distress poor sleepers functioned similarly to the high-distress good sleepers. This seems to give some insight into the "dysfunction" seen in some insomnia patients. The dysfunction is more tied to the patient's view of her sleep quality (and the resulting stress over it) than it is to her actual sleep!

So now we come to the first tough question I'm going to ask you about your sleep problem:

Could your distress about your sleep be, at least in part, contributing to your problem? Is this sleep thing a bigger deal in your mind than it is in reality?

What we need here is some honest feedback about how others perceive you and the magnitude of the role your sleep issues play in your life.

 SLEEP DISTRESS EXERCISE

1. Find a friend you know well, but with whom you are not in a relationship.
2. Tell this friend you are doing a project for an online class.
3. Outline the project as being an exercise in examining the things that set individuals apart from others.
4. Tell your friend that you are going to read some questions and your friend is to say yes or no to the queries.
5. For the first question, ask, "Am I a good person?" Your friend will say yes, and the question will relax him. If your friend responds no, I can see why you might be distressed at night when you are trying to sleep.
6. For the next questions, ask, "Am I good at my job?" and then "Am I reasonably healthy?"
7. The next question is "Do I sleep well?"

   Outside of living or sleeping with someone, this is not a question a friend should be able to answer easily *unless he hears you talking about it.* I've known my assistant, Tammy, for ten years. We work in a sleep clinic where we talk about sleep for a living. I have no earthly idea how well she sleeps. I assume it's pretty good because she never complains.

8. If the answer is any of the following, you may have a problem:

- No.
- God, no!
- Laughter, followed by "Are you serious?"

At this point, refer back to the answers to the step 6 questions. If the answers to those questions are yes and yes, you may be more distressed about your sleep than you think.

Insomnia Identity

In my opinion, Charles Morin's research on insomnia and writing on insomnia are gospel. His 1993 book, *Insomnia: Psychological Assessment and Management,* is a sacred text to those in my profession. Through his work, Morin really created an Old Testament for the treatment of insomnia, which virtually anyone with sleep problems can recite. Like Moses delivered the Ten Commandments, so too did Morin spread the word and attempt to deliver his people to the promised land of sleep. Here is how I would boil down the essential messages of his work.

THE TEN SLEEP COMMANDMENTS

1. Thou shall not worship any sleep aids, noise machines, or iPhone sleep apps.
2. Thou shall not create any graven image of insomnia and then blame it for all of the bad things in thy life.
3. Thou shall not take the name of the Lord in vain when thou triest to sleep but cannot.
4. Remember the Sabbath. Keep it holy. Quit sleeping in on these days.

5. Honor thy mother and father. Quit blaming their genes for your insomnia.
6. Thou shall not kill, steal, or commit adultery. The guilt will really screw up your sleep.
7. Sleep is the most important thing in the world. Tonight's sleep is relatively meaningless.
8. Thou will not be as bad off as thou thinkest you will be tomorrow after a bad night.
9. Use the bed for sleep and sex only. If thou are in bed not sleeping or procreating, leave.
10. Thou shall not covet your partner's amazing sleep. You'll never sleep like that—lower your expectations.

Okay, so these are not *exactly* Morin's pointers about sleep, but they might as well be blessed commandments because most of what you "know" about your good sleep habits is from Morin. My guess is that the commandments are not new to you at all. If reading "use the bed for sleep and sex only" cured you, (1) you're welcome and (2) what rock have you been living under? These standard tips appear in every sleep book, magazine article, and blog on the planet.

Because of the chronic nature of patients' insomnia, many begin to incorporate their sleep issues as part of their core identity. In some cases, being a "bad sleeper" becomes central to who they are. The problem with that is when someone abruptly challenges something at the core of who you think you are, the consequences can be severe.[66] Look at the devastation caused by divorce. Suddenly, you are no longer a husband or wife, and this central defining aspect of your identity is gone.

If you are reading this section, consider openly that there might be a teeny tiny hint of truth in what I'm saying. Does your family know you struggle to sleep? If so, why? Was it in your yearly holiday card?[67] Do you tell people at parties whom you've just met that you

---

66 Think Luke Skywalker's reaction to Darth Vader's paternity announcement.

67 Is it *always* in your yearly holiday card? "Merry Christmas! Greetings from Times Square! I'm still not sleeping. Hopefully Santa will finally bring me the gift of sleep this year. Come visit me!"

can't sleep? If someone says that they struggle to sleep, do you feel an overwhelming urge to top their sleepless story like that Kristen Wiig character on *Saturday Night Live* who is always just a bit better than everyone else?

**See insomnia for what it really is and see yourself
for who you are.**

Here we go again. I'm going to say something that is going to make you mad and throw my book away.

Insomnia is not that big of a deal.

Take a look at the top 100 causes of death:

cardiovascular disease, cancer (all kinds), respiratory disease, unintentional injuries, stroke, Alzheimer's disease, diabetes mellitus, respiratory infection (influenza, pneumonia), nephritis/nephropathy, suicide, blood poisoning, liver disease, hypertensive heart disease, Parkinson's disease, homicide, infectious/parasitic disease, heart attack, HIV/AIDS, chronic obstructive pulmonary disease, perinatal disease, digestive disease, diarrhea diseases, gun violence, war, tuberculosis, malaria, lung cancer, road traffic accidents, childhood diseases, neuropsychiatric disorders, stomach cancer, diseases of the genitourinary system, cirrhosis of the liver, colorectal cancer, liver cancer, measles, maternal conditions, congenital malformations, nutritional deficiencies, breast cancer, esophageal cancer, inflammatory heart disease, dementia excluding Alzheimer's, falls, drowning, poisoning, lymphoma/multiple myeloma, rheumatic heart disease, oral/oropharynx cancers, fires, pertussis, prostate cancer, leukemia, peptic ulcer disease, protein-energy malnutrition, endocrine disorders, asthma, cervical cancer, pancreatic cancer, tetanus, sexually transmitted diseases, bladder cancer, meningitis, syphilis, neoplasms that are not malignant, iron deficiency anemia, ovarian cancer, tropical diseases excluding malaria, epilepsy, musculoskeletal diseases, hepatitis B, alcohol use disorders, drug use disorders, uterine cancer, skin diseases, melanoma and other skin cancers, hepatitis C, leishmaniasis, trypanosomiasis (African sleeping sickness) . . .

I give up. That's the most complete list I can put together. Notice any diagnosis missing from that list? Exactly. Insomnia.

Nobody dies from insomnia. You're fine. You are more likely to die of too much sleep (African sleeping sickness) than of insomnia. Stop worrying about it so much.

Brilliant sleep specialist Michael Thorpy referenced this in a *New York Times* blog post titled "Can You Die of Insomnia?" In it, he stressed to readers that sleep deprivation was different from insomnia, and while chronic insomnia will not directly lead to death, sleep deprivation increased one's risk for developing other serious medical conditions that do have increased mortality. These conditions are not the same thing. It would be great if someone could pass that message along to the media at large.

Insomnia ≠ Sleep Deprivation

These terms are not synonyms. Please divorce them in your head. When Matt Lauer is on the television talking about how sleep deprivation was the cause of a train crash in the state of New York or how a study of sleep-deprived shift workers showed they are predisposed to some bad health outcomes, he is not talking about you. Why is this important? Because not understanding this produces the single most important ingredient when it comes to insomnia: fear.

Insomnia Is Fear

I've spent more than twenty years of my life working with and thinking about sleep. Spending that amount of time in a field of study in a subject as small as sleep has afforded me the chance to see thousands of patients, be mentored by many incredibly intelligent sleep physicians, and read the research and thoughts of countless others. Like the way Stephen Hawking tried to consolidate the workings of the universe into a single unified theory, my pedestrian brain has tried to distill the complicated field of insomnia into something much simpler: one word.

For me, I feel that insomnia is really all about one thing: fear.

When I was young, I was not a huge fan of the dark. One night a friend of mine and I tried sleeping outside in a tiny little house my father had built for us in the woods. I remember lying in my sleeping bag, looking at my friend and thinking, "This is never going to happen." I literally recall listening to the introduction of Gary Wright's "Dream Weaver" playing on my little battery-powered JVC jambox and basically losing it. We both went racing back up the hill to my house and the security of my early-1980s bedroom.

So let's break down the situation further. We showed up to sleep in the little house. By *little house*, I mean it was literally a little house: 2×4 construction with insulation, wood siding, roof shingles, and so on. It even had its own little deck overlooking the woods behind our house. It was well built and totally secure. It locked from the inside. Nothing could get inside that house. I would wager an angry bear would have had no success breaking in. In other words, there was nothing rational we had to fear.

Fear like that is not rational though. It's like a fear of clowns (which, incidentally, were featured on the wallpaper of the tiny house). We feared something that had no real grounds for being feared. That is what is so magical about fears. They have absolutely no dependence on logic or reality. They can be given a reality (such as "We were down there and we heard bear footsteps and narrowly escaped with our lives."), but usually this is just a construction meant to validate, not explain.

So there we were, back in my bedroom. "Dream Weaver" had mercifully been replaced by "Centerfold" by the J. Geils Band. No bears or deranged escaped-prisoner murderers loose in the woods could be seen or suspected in my room. . . . We were asleep in minutes.

The truth was that I made the decision not to sleep well before we got into our sleeping bags that night in the little house. As we made a list of all the things we would need for the overnight adventure, I kinda knew in the back of my mind we would end up in my bedroom. I was more than willing to give it a try, but my mind was made up before we even left the house that I probably would not sleep. Once I got outside and felt the struggle begin, the fear of not sleeping set in and that was it.

Fear is a central component of insomnia. For something like insomnia to have any power over you, fear must play a primary role. Oh, you can call it different things, but no matter how you slice it, the patients in my office are there, in part at least, because of fear.

"I'm *concerned* that if I don't sleep, my health will suffer."

"I *dread* the loneliness and boredom of being awake during the night."

"I *worry* that I'll be dysfunctional at work the next day and I can't afford to not be sharp in my line of work."

"I *discover* that my other health problems and pain I experience are worse when I don't sleep, so my rheumatologist says it is especially important for me to get my sleep at night."

Underscoring all of these statements is fear. For each italicized word, substitute the word *fear* and you can easily see another motivation behind the statement. Patients get pulled into this fear, but so too do other family members, physicians, and other practitioners. Think about it. When your kid says, "Mom, I can't sleep. . . . I haven't slept in weeks," what would your response be?

You will never manage your insomnia without gaining control and perspective over the fear you can experience when sleep is not happening in the way in which you want it to. Imagine going to bed tonight and finding yourself awake after thirty minutes. Wide awake. What will the dialogue in your head sound like? What if it is totally illogical that you are awake?

I remember back in my medical school days in Atlanta that I would sometimes get up early and head off to class, where I would be all day. I'd meet up with my wife after classes were over and she was done with teaching for the day. We would go to the campus gym and work out, go home, fix dinner, and after eating I would go back to the sleep center and work at night running research studies. I would then be awake all night and return home Saturday morning. Even after the countless hours of being awake and the state of total exhaustion I was in, I can vividly remember climbing into the futon bed and sometimes struggling to sleep. "This is really weird," I can remember thinking, being amazed at my brain's ability to be a real ass about sleep at that moment.

But what's most important is, I remember not really caring. The bedsheets were cool and comfortable. The room was dark and quiet. I did not have a pathophysiology book in front of me or a stack of bills to pay. I was just there, drunk with sleepiness, yet wide awake. Did I care? Not at all. Did I fear the consequences? No. I thought of it as kind of a win-win. If I fall asleep: win. If I don't: win. At least I don't have to go grocery shopping.

When you talk with good sleepers, they all have a flip-flops-and-Hacky-Sack mentality toward their sleep. "Whatever, dude." Within them is an inner belief that they are basically going to be okay no matter what happens that night in bed. This is the mentality you must find, or you will be doomed to struggle forever.

Fear about "not sleeping" is everywhere out there. Recognize it. Do not fall into the trap. Control what you can control. After that, forget about it. I know it is a hard thing to do. You've been struggling to sleep for so long. But you can do it.

The Painful Truth:
Understanding Primary Insomnia

Sadly, I must admit that for a small group of people, they cannot seem to get their difficulty sleeping under control.

They have tried everything. Books upon books. Internet intervention and hypnotherapy courses. Doctors, specialists, therapists . . . nothing. This is beyond hard insomnia. This is malignant.

Many books chicken out when it comes to individuals who have chronic, impossible-to-treat, "pills don't work on me" insomnia. They call it things like "primary insomnia," suggest sufferers tidy up their sleep hygiene and spend less time trying to get sleep, and move on. Sometimes a novel pill is suggested, but it's usually accompanied with a shoulder shrug and a patronizing pat on the head.

It is interesting that many insomnia books end with primary insomnia in much the same way Jimmy Kimmel ends his show, stating, "Sorry. Our show went long and we had to bump Matt Damon. Hopefully he can join us tomorrow." Jimmy doesn't really intend to talk to Matt, and many sleep doctors who write books don't intend to talk about primary insomnia.

What is primary insomnia? Hell if I know. Hell if anyone knows. I could write something to the effect of "It's when the brain does not make the chemicals necessary for initiating and maintaining sleep," but I'm not sure I or most other sleep doctors really believe that is the case in most instances. I think people out there with primary insomnia exist, but they are as rare as Milwaukee Brewers World Series rings.

Whenever I see someone I think might just be the real deal, my evaluation of her sleep either through logs, or actigraphy devices, or a genuine sleep study usually disproves my hunch. Beyond that, her "disability" is usually nonexistent, and whatever that disability is, it virtually never involves excessive daytime sleepiness. It's been many, many years, and I have not found someone yet.

In other words, for all of the horrible things that such a patient's insomnia is causing, it seems to be doing a pretty fine job of making the patient feel awake during the day. Really awake in many cases.

Consider that carefully.

The problem is, even if someone with primary insomnia does appear in my clinic, the truth is that sleep science does not have a solution. We can run through all of the sleep aids detailed in the next chapter. We can try the antidepressants. We can use more unusual drugs like sodium oxybate, a narcolepsy medication similar to gamma-hydroxybutyrate (GHB).

The bottom line: The ugly truth is that if you have true primary insomnia, current sleep medicine may not be able to help you. You might be saddled with insomnia for the rest of your life. My best advice to you is to work on cultivating an attitude of acceptance. The condition is not fatal. In fact, as we have already seen, your attitude about the sleep problem may play a huge role in your ability to function at a high level. Look at it from a positive perspective: The condition frees more time for you to do things in the evening. Yes, you might feel a bit fatigued, but there are medications to treat that too if you like.

In sports, I was always taught to control what I could control. I try to teach my kids the same thing. Sorry if this sounds preachy, but you can't control whether or not your insomnia therapies work.

You can control only your response to the delays you experience trying to fall asleep.

This leads to my plan: Fake it till you make it. As of right now, you are a great sleeper. On the nights you sleep like a star, you're not surprised. On the nights when your sleep is a little less than fantastic, it's no big deal . . . a bump in the road.

 ## CONTROLLING YOUR MIND EXERCISE

1. For one entire month do not talk about your sleep. Don't talk about it at all. If you are asked directly about your sleep, respond with a simple "I slept fine." Not talking about your sleep includes not blaming something that happens or something you feel on poor sleep. "I'm sorry, team, for struggling a little this morning . . . I had a rough night" is a no-no.

2. For one month, avoid any exposure to media related to sleep. This means self-help books (but finish this one, obviously), Internet sites/blogs, television shows, magazine articles, and the rest.

3. If somebody were to ask you what time you typically fall asleep (not go to bed), make that time the earliest you get into bed at night.

4. Practice a goal-directed activity when you are in bed awake. Use the time for meditation. Work on clearing your mind and relaxing your body. Don't let stress in. For many patients, simply resting can be remarkably restorative. Make rest (something you can control) the goal, not sleep.

5. Another strategy is to imagine yourself performing a task at night. I give my athlete patients tasks related to what they do. For a basketball player, I say, "I want you to shoot fifty free throws perfectly." For a pitcher: "I want fifty perfect pitches." I've got a patient who likes to play golf while his wife likes to imagine baking banana bread. Regardless of what you pick, you have

to imagine every detail, right down to the bruise on the banana you are peeling. Because the brain cannot easily tell the difference between imagining an activity and actually doing it, you might find your golf swing is a bit improved as well as your sleep satisfaction.

6. At some point in the day, make time to think about the fact that you are a good sleeper. If you ever have the chance, take a picture of your feet in a hammock (or simply find someone else's picture of feet in a hammock). Post it on your Instagram account and write a caption, "Nothing like the beach and a hammock to make you sleep like a baby." Hey, fake it until you make it.

7. When all else fails, join the army or complete a medical residency. Nobody has trouble sleeping in boot camp or while on call in the hospital!

Final point: remember my endless list of the causes of fatigue back in Chapter 3? It is very easy to acquire one of these conditions and start to feel really fatigued. Day after day of waking up feeling like you don't have the energy to push the power button on your television remote can make you start to blame your sleep for your devastating fatigue. (This can be strongly hinted at if your Epworth Sleepiness Score is less than 10!) As their fatigue worsens, patients begin to stress more and more about their sleep because they have decided it is why they feel poorly during the day. They go to bed early to get more sleep, only worsening their ability to fall asleep.

Keep in mind, while dysfunctional sleep can certainly lead to feeling poorly, it is usually associated with an *increased* drive to sleep, not the *decreased* drive many insomnia patients have. Investigate your sleep, but you and your primary care doctor should never put all of your eggs in the sleep basket; it may prevent you from discovering the true cause for why you feel like hell.

That's all I've got. It's pretty much all anybody's got. If you are not a perfect sleeper yet, stick with it. People are rarely perfect cyclists the first time they get onto a bicycle. Sleep is a skill. You can improve it. Maybe you can perfect it.

CHAPTER 10 SUMMARY

1. Fear and helplessness are the fuel that powers insomnia. Education about sleep is the solution. Controlling what you can control and letting go of the rest is the solution to your helplessness. You have power here. Insomnia can exist only in an individual who cares.

2. Sometimes you control everything you could possibly control, you get into bed after being up a really long time and you still don't fall asleep immediately. Sometimes Appalachian State comes into Ann Arbor and beats Michigan in football. Sometimes Buster Douglas beats Tyson. Sometimes the USA beats Russia in Olympic hockey. I can't explain it. It happens. Don't get too upset about it and move on.

3. Control your fear and anxiety about the situation. If sleep is not happening, just relax and enjoy the peaceful time. Resting helps your body too.

4. Sleep disorders are only one of many things that make people feel bad during the day. Investigate all the possibilities. Don't get hung up on your sleep. It may not be all that bad.

I know what you are thinking. "For the love of God, have mercy on me and just drug me." Sleeping pills are like tigers: I'm not sure either is really suitable to keep in your home long term. Unlike tigers, sleeping pills are everywhere, so let's dive into them.

SLEEPING AIDS
The Promise of Perfect Sleep in a Little Plastic Bottle

N 2015, **KAREN WEINTRAUB WROTE** a short article for the *New York Times* titled "Do Sleeping Pills Induce Restorative Sleep?" That's an intriguing question, which we will get to later. What interests me the most about this article and scores of similar articles are the throwaway comments made as the topic is introduced. In the article she writes, "There is quite a bit of evidence about the negative health consequences of insomnia, but researchers don't know precisely what it is in the brain and body that is 'restored' by sleep to aid optimal function."

The point of this sentence is that people like me don't know exactly what magic happens when we sleep to make us feel great the next day versus feeling like a truck ran over us. The problem is that this reporter (and countless others) writes things about the "negative health consequences of insomnia." Is there really quite a bit of research demonstrating the negative health consequences of *insomnia,* or is what Weintraub is meaning to say is there is quite a bit of evidence about the negative health consequences of *lack of sleep*? Do you see what happened there? The author is using the term *insomnia* interchangeably with the term *lack of sleep,* and I don't need to tell you, now that you've read nearly two-thirds of this book, that the two are not one and the same.[68]

Take This Pill or Else

Before we jump into the topic of sleeping pills and sleep, I want to make one point very clear. I'm not a big fan of sleeping pills. I'm going to do my best to make you less of a fan as well. If I succeed, I do not want you to adjust the way you take these pills or discontinue these pills by yourself. Instead, dear Reader, I want you to have a discussion with the person who prescribes your medication. Abruptly stopping some sleeping pills can carry risks. I don't want bad things to happen to you. I want good things to happen to you, and I want you to help educate your prescriber. That happens with dialogue. So basically, don't go rogue with your sleeping pill use, okay? Great!

As we begin to look at the incredibly rampant use of sleeping aids, we need to understand the motivations for this behavior. Most people and patients I meet do not like pills. "I'm not a pill person" is a common phrase in my clinic. People do not like being on medication. Old people, sick people, drug addicts take pills. Healthy people do not. People love to tell me they are on only a half dose. There is also a perception that drug companies are evil and that their pills are part of a conspiracy to get us hooked on them so they can make big bucks and use the money to influence doctors with fancy ballpoint pens. Moreover, people these days are just more skeptical about the substances they put in their body. They want free-range, cage-free, organic drugs that are produced in small batches at boutique lavender farms. They don't want "chemicals" in their body.

So why on earth are people so eagerly agreeing to take sleeping pills for their insomnia? Because they don't want to die of a vicious blend of heart attack, stroke, and dementia. Put another way, they want no part of the "negative health consequences" that everyone knows are associated with insomnia. Despite not wanting to take the pills, even more they don't want to die!

---

68 Remember: Insomnia ≠ Sleep Deprivation

This brings us to the mixed blessings of the scientific process and the overwhelming presence of the media, which do not always get it right. Here is an example of what I mean provided by Karen Johnston, who was my neurology residency director and is now chair of neurology at the University of Virginia. She often gave the fictional example of a research study looking at individuals who carry matches in their pocket. This study concludes that these individuals have far more risk of developing lung cancer than non-match carriers. The conclusion and the message: Carrying matches causes lung cancer.

Well, not really. There are some very important details missing from the equation, the biggest being match carriers probably smoke cigarettes. These details matter and they matter with insomnia. The "health consequences" of insomnia are abstract, often predominantly psychological and poorly defined. The health consequences of inadequate sleep are very clear and very serious. Because the media do not really understand the difference between a typical insomnia patient who "can't sleep" and scores a 1 on an Epworth Sleepiness Scale and the typical night-shift worker who works a second job and is routinely sleep deprived to the point where he is falling asleep while on the toilet, they use the two situations interchangeably. This steady pipeline of reporting the devastating effects of true sleep deprivation under the umbrella of "insomnia" is making consumers feel like they have no choice in the matter. Take a pill to sleep or die.

Don't believe me? Spend a week in my clinic. Sit down and have a chat with a twenty-year-old recent college graduate who had a full-on panic attack complete with tingling lips, numb hands, and shallow breathing when I said, "My goal is to get you off the big doses of Ambien you've been taking over the past few years and . . . dear, are you okay?"

So the messages are out there: Sleep for optimal health. Get your eight hours a night or face the consequences. Lack of sleep makes you fat. Lack of sleep leads to heart failure. Lack of sleep may lead to breast cancer. In the face of all these dire warnings, what conclusion would any rational individual reach after a night of poor sleep? *I better get my sleep on fast or I'm toast.*

Enter the sleeping pill to save the day. Most people see a television advertisement for a sleep aid as an offer of help. In reality, these ads continue to reinforce the idea that lack of sleep is devastating to health and insomnia patients are losing sleep. They imply medications are necessary for individuals to fall and stay asleep and that this is the simple, safe, and only solution for the problem. It's also nice to know you're not alone. There are attractive middle-aged people in all kinds of lovely bedrooms across this great land struggling to sleep.

The problem is, the promise of these pills is a bit empty. I have never read a study that has shown these pills to reduce the time it takes to fall asleep by more than a few minutes, nor add more than a few minutes of total sleep to the user's night.

Make no mistake, these sleeping pill ads are not new. Pharmaceutical companies have been helping facilitate the idea that their drugs solve sleep problems for years. How convenient to sell a product for a condition that has so little real downside . . . like a pill that prevents you from not wanting to eat lunch occasionally.

And it doesn't stop with adults. Man, the stuff that we throw at kids to help them sleep is enough for its own book. Books you have to read in a whisper, sleep-training lights, roll-ons with interesting ingredients, and more. Not only is this process completely unnecessary but it sets up a whole new generation of patients growing up feeling like they cannot sleep and need pills and medication to do it.

Media: Everyone Is Taking Sleeping Pills. It's Fun!

There is a tremendous media machine that is churning out fear and misinformation about sleep. It comes in the form of funny television characters, like Karen Walker from *Will & Grace*, who embody what is portrayed as a modern approach to sleep. Reading between the lines, the viewer is told, "Nobody struggles with insomnia anymore. We just take a pill," leaving the viewer to feel a little dumb for actually trying to just get into bed and fall asleep without help.

In countless episodes, Karen would announce to anyone who was around how essential her alcohol or medication consumption was to her sleep. Among her more memorable lines: "Normally my motto is 'Drugs not hugs'"; "Plus, I'm high most of the time, so there's that"; and "I may be a pill-popping, jet-fuel-sniffing, gin-soaked narcissist . . ." This character is clearly over-the-top. In one episode, she uses her Valium pills and other medications as color swatches to help an expecting couple plan out how to paint their new home.

Please don't paint me as someone who fears medications. I'm not. By the same token, please don't paint me as someone who did not immensely enjoy all eight seasons of *Will & Grace*, because I did. The casual way, even helpful way, we portray sleeping pills in this country via television characters, pharmaceutical ads, and other popular culture references is very problematic though. It provides the lure of an easy way to achieve good sleep, which has the unintended consequence of making the effort it often takes to achieve healthful sleep seem unnecessarily hard. "No, thanks with the sleep diary, the A.M. exercise, and the sleep restriction. . . . I'll just take some of the sleep doctor pills you must keep in the back of your office, and I'll be on my way." Sedation and sleep are not always the same thing.

Managed Care: No Time to Treat Sleep

Do you want to know the other big reason people are so into sleeping pills like they're free Barnes & Noble Wi-Fi? It's because the current economics of medical care in this country don't allow doctors enough time to fully address the needs of each patient. Because primary care doctors can't add hours to a day and insurance companies frequently reduce what they can charge per patient visit, the time doctors can spend with individual patients keeps shrinking and shrinking, and guess what gets squeezed out. Sleep. The primary care doctor acts with a sort of triage mentality. Blood pressure and diabetes are at the top of the list, with obesity and cholesterol issues right up there. After dealing with these heavyweight[69] issues, not much time is left to chat about sleep problems

---

69 No pun intended.

with a waiting room full of patients . . . waiting. What is a doctor to do? Reach for a sleeping pill and cross his fingers.

Sleeping pills were designed for sporadic use, and in some appropriate cases they work very well. They were not designed to sedate individuals to sleep on a nightly basis. Think about the food analogy again. How many times have you sat down for a meal and not felt particularly hungry? What did you do? Me, I immediately panic about the effects of malnutrition on my body and seek out appetite-stimulant medications and take them so that I can artificially make myself hungry and choke down some food. Of course, eating like that makes it difficult for me to feel hungry when the next meal rolls around, so I just keep taking more and more of the pills. Otherwise, I'm doing great.

This sounds ridiculous, and it is. If you are not hungry for lunch, you skip it; it's no big deal. Why is it then that when a patient relays to a doctor that she is having trouble sleeping at night, the patient is usually given pills? If your response is "Because the patient can't sleep. This patient is not just an every-now-and-then insomnia patient. This is a really bad insomnia patient and if something is not done, the patient could die," you are in fact wrong. Repeat this 100 times to yourself: Everyone sleeps. Remember those primary drives from Chapter 2?

With no time to educate patients and no time to listen, a difficult situation is often established when doctors treat insomnia. The patient is upset. The patient is desperate for help, understanding, and compassion. The doctor, however, is over an hour behind schedule and really does not have the time for an in-depth discussion about what we have covered in this book. What he does have time for is to pull out another prescription and write "Ambien 10 mg." He's happy because he gets to move on with his day. The patient is happy because pills always work. Each party departs hoping that the insomnia just goes away.

But it doesn't. The patient will invariably return and the doctor will invariably write the prescription because the patient has "gotten so that I can't sleep without it." Without knowing it, this Dr. Smith has become Dr. Frankenstein . . . because he's created a monster.

Fifteen years later, Frankenstein's monster comes crashing into my office in a panic because Dr. Frankenstein has enraged the monster not only by refusing to create a mate but also by refusing to refill the monster's prescription because of either concerns about addiction or the growing body of evidence potentially linking some of these pills to memory loss, confusion, or even dementia with prolonged use. Believe me, it is far easier to deal with insomnia at its beginning (simple insomnia) than at this point in the evolution.

If insomnia were a relatively uncommon complaint, you could justify this treatment pathway, but it is not. Insomnia is always in the top ten of primary care patient complaints, yet given what we know about sleep treatment within a primary care setting, it is often largely ignored outside of throwing sleeping pills at the situation. Here are the top patient issues:

abdominal pain	headache
back pain	*insomnia*
chest pain	numbness
dizziness	shortness of breath
fatigue	swelling

It's no surprise that pain is well represented. It's usually number one. So working on that assumption, we can distill this list down to seven items, with *pain* covering abdominal pain, back pain, chest pain, and headache.

pain	numbness
dizziness	shortness of breath
fatigue	swelling
insomnia	

Once again, we see the word *fatigue* in the mix. Patients will generally use terms like *fatigue* and *sleepiness* interchangeably, so now we have the two main categories of sleep medicine represented within the most common complaints seen by primary care doctors: "I can't sleep" and "I'm too sleepy."

Sleep patients—including insomnia patients—make up a big population of people coming to see primary care doctors. These docs need to embrace the whole of sleep medicine and stop thinking of pills as a meaningful long-term solution.

And in fairness, I think they are. I'm seeing more and more doctors counseling their patients on the dangers of sleeping pills. Their addictive potential is discussed. Doctors are becoming more aware of counseling strategies and CBT-I. At the very least, primary care physicians are establishing limits and boundaries. When their patients want to double their Ativan to help with their sleep initiation, doctors are beginning to say, "I think your sleep issues are moving past my expertise and comfort level. I want you to see a sleep specialist." Hallelujah!

Types of Sleeping Pills

Before we talk about situations in which sleeping pills can be useful and appropriate for patients, I think understanding how sleeping pills of different kinds work can help you better understand which, if any, is right for you.

Over-the-Counter Sleeping Pills

It is every pill's dream one day to hit it big and get its own section in CVS pharmacy. Sleeping pills have indeed hit it big, with shelves upon shelves of sleeping pills available to choose from as you wait for your prescriptions to be filled.

I recently stopped into my local big-chain pharmacy to get a firsthand look at all of the various options consumers have to choose from. If you are looking for variety in terms of over-the-counter sleeping pills, you are out of luck. Despite the incredible array of colorful boxes, generic knockoffs, and two-for-one sales, the active ingredients in over-the-counter sleeping pills are all basically the same: antihistamines.

Remember histamine from Chapter 5?[70] Its chemical structure looked like sperm. Ah, now you remember. Histamine makes us

---

70 If you take lots of antihistamines you may not remember because they can affect memory.

feel awake and alert. Blocking histamine with an antihistamine has the effect you think it would. It makes you sleepy. Do antihistamines work? Yes, but they are not particularly strong. In the elderly, these drugs can have some negative side effects like memory loss and confusion the next day, so use them with care.

Melatonin

Wow, let's go way back to Chapter 3 and melatonin. Melatonin, as you know, is the "chemical of light." Some people take melatonin to help them sleep. It's popular among pediatricians for helping children sleep. I'm not exactly sure why. My guess is that it is perceived as generally harmless.

The drug seems to be most beneficial for helping with circadian problems like jet lag. As a long-term sedative, its effectiveness is questionable.

 CUTTING-EDGE SCIENCE

THESE DAYS EVERYONE USES MELATONIN. It's—how can I say it?—*en vogue*. Does it really work as a sleep aid? A 2014 study showed the effectiveness of melatonin to prevent jet lag and promote sedation to be "weak." The study was very thorough and seems to lead to the conclusion that melatonin is probably about as effective but also about as harmful as sleeping turned 180 degrees in your bed. Think about this study if you feel like you can't sleep without your melatonin.

Valium and the Benzodiazepine Gang

In 1955 scientist Leo Sternbach accidentally synthesized the first benzodiazepine, chlordiazepoxide, a precursor to Valium. The drug became very popular around the world as people took it and immediately stopped caring about things. The sudden popularity of Valium among housewives inspired the Rolling Stones song "Mother's Little Helper," a reference to the drug.

These tranquilizers quickly became adopted for use in seizure control, muscle relaxation, anxiety control, and sleep assistance. While generally safe from a tolerability perspective, their sedating properties combined with their addictiveness led to some bad outcomes, particularly when they were mixed with alcohol and other sedating drugs. Recently, reports linking these drugs to cognitive decline later in life have emerged, making these pills much less savory and more than a little scary. But the "old guard" docs still use them like crazy.[71] This would be a great time to check your grandmother's medication list and see if any of these drugs are on it. If you find some, you might want to make her an appointment with a new doctor.

Along with Valium (diazepam), there are a lot of medications in the benzodiazepine family: alprazolam (Xanax), clonazepam (Klonopin), estazolam (ProSom), flurazepam (Dalmane), lorazepam (Ativan), midazolam (Versed), temazepam (Restoril), and triazolam (Halcion).

These drugs also have been shown to suppress slow-wave sleep. This is an unfortunate outcome for someone wanting to feel better the next day. This is where sedation and sleep deviate. Remember, for sleep to have an impact in your life, it needs to include a robust amount of deep sleep and all of the restorative things that go along with it. This is sleep. Simply sedating someone does not produce this effect. While nobody knows exactly what led to the death of Elvis Presley, Valium was a drug commonly linked to The King and widely thought to have contributed to his early passing. Sedation, at times, can be dangerous and should never be confused with sleep.

Ambien and His Imidazopyridine Buddies

Despite the fun everyone was having with benzodiazepines interfering with their ability to breathe on Saturday night when too many were taken with their wine, the search was on for a new, safer

---

71 Every community has one of these doctors practicing medicine. He usually has white hair and a name that went out of style when prohibition was repealed, like Jebediah, Alastair, or maybe Mathias. He's skeptical about handwashing, hates computers (sometimes uses the word *machine* or *robot* interchangeably with *computer*), and is unaware of any drug that came out after the Carter administration. This doctor will give benzodiazepines for anything from a runny nose to jock itch.

drug. Enter Ambien (zolpidem), a drug introduced in 1993. This drug was a miracle, as it seemed to feature only a sleep-promoting effect without all the dirtiness associated with benzodiazepine. The world would be saved because insomnia would surely be eradicated like smallpox!

Unfortunately, and we're still not sure what happened, this was not the case. Despite this new drug, people continued to have insomnia. What's more, users of this drug started to do some really bizarre things at night. Acting out dreams, eating food with no memory of doing so the next day, even driving and having sex. These things have led to stricter control and stronger warnings, particularly in women.

No problem, there are other drugs that work like Ambien to choose from. ZolpiMist is basically Ambien nasal spray. Intermezzo is a smaller dose of zolpidem designed for individuals who wake up during the night and can't get back to sleep. Ambien CR is a longer-acting formula for when you need more Ambien. I honestly have no idea who uses Ambien CR and why. More is not always the answer. Just go see a sleep specialist already! The company that makes the drug warns users that you should not drive after using Ambien CR. Don't believe me? This is straight from the package insert that goes with the drug:

5.1 CNS Depressant Effects and Next-Day Impairment

AMBIEN CR is a central nervous system (CNS) depressant and can impair daytime function in some patients even when used as prescribed. Prescribers should monitor for excess depressant effects, but impairment can occur in the absence of subjective symptoms, and may not be reliably detected by ordinary clinical exam (i.e., less than formal psychomotor testing). While pharmacodynamic tolerance or adaptation to some adverse depressant effects of AMBIEN CR may develop, patients using AMBIEN CR should be cautioned against driving or engaging in other hazardous activities or activities requiring complete mental alertness the day after use.

Patients with insomnia are tough to treat and they do not give up their sleeping pills without a fight. One of their favorite excuses is that they must take the Ambien in order to get sleep so that they can do their work and not lose their job. But even the most pill-

174 THE SLEEP SOLUTION

happy insomniac needs to recognize that if his job involves getting up and driving to said job, taking this drug really is not compatible with that activity.

The pharmaceutical industry did not stop with Ambien. Sonata (zaleplon) has a really short half-life, so it is often used by people who have difficulty falling asleep or wake up in the night and do not have time to take a longer-acting drug before they have to be up and driving in the morning. But there are warnings against "sleep-driving" and driving in general the following day with this drug too.

Lunesta (Eszopiclone)

Lunesta is another nonbenzodiazepine that is in the cyclopyrrolones family of drugs. Lunesta is the only one commercially available in the United States. It comes in 1, 2, and 3 milligram doses so that the user can pick his desired strength. Usually 1 milligram doses are for simple sleep-onset issues and 3 milligram doses are more for sleep-maintenance issues or early-morning awakenings.

Rozerem (Ramelteon)

In 2005, Rozerem made a big splash when it was approved to treat insomnia. Unlike the benzodiazepines and nonbenzodiazepines, this was the first drug that did not target GABA, an inhibitory neurotransmitter in the brain, to produce its sedating effects. Instead, this drug works at the melatonin receptor. It also has the distinction of being the first drug approved for long-term use. Studies aside, the drug has never made much of a splash; many users were fairly unimpressed by its effects. Summary: meh.

Suvorexant (Belsomra)

Want something new and shiny? Suvorexant is exactly what you need. This drug received its approval for insomnia in 2014 and acts as an orexin (makes you awake) receptor antagonist, which basically means the drug prevents orexin from making you feel awake. Overall, the dosing is low and the effects are considered fairly tame. Because it affects the same neurotransmitter that is deficient in narcolepsy, studies of the drug show that some of the unusual symptoms of narcolepsy, like sleep paralysis and cataplexy (suddenly feeling paralyzed), can occur with this drug. The description of these side effects during the Belsomra commercial simultaneously horrified my wife and wildly amused my kids.

Silenor (Doxepin)

Silenor is a tricyclic antidepressant that is often used to treat insomnia. Other tricyclics, such as amitriptyline, are often used as well. These drugs have been around for a while. Doxepin was introduced in 1969, and amitriptyline was introduced in 1961. These drugs can aggravate restless legs syndrome in some people.

Antidepressants/Antipsychotics (and Other Drugs That Have No Business Being Used as Sleep Aids)

Quick, guess which prescribed medication is the most common sleep medication. Time's up. Here's a hint: It's an FDA-approved antidepressant with no FDA approval for sleep. Give up? It's trazadone. Trazadone is just one from a long line of off-label antidepressants often used for sleep. Another is Remeron. The great thing about Remeron is that its name implies that taking it will turn REM on. Unfortunately, it all too often turns on weight gain too.

Bored with antidepressants? I am. The new fun thing to do is to skip the antidepressants and go right to the antipsychotics. Drugs like Seroquel (quetiapine), Zyprexa, and Risperdal (risperidone), which used to be used exclusively for patients with mania or psychosis, are now being used off label to treat simple insomnia. In fact, there is an emerging opinion that the benefits of these drugs do not outweigh their risks in the treatment of sleep disorders. There is no real literature supporting their off-label use in helping patients fall asleep faster or even stay asleep. To me, these drugs epitomize the haphazard, misinformed, and dangerous practices of some doctors who do not understand sleep or how to treat it.

I guess I'll go ahead and throw propofol in here too, as I know of at least one doctor who used it to help his patient sleep. That patient was Michael Jackson, who died from his doctor's ignorance, just as my patient with an aortic aneurysm would die if she asked me to fix it. I'm not a cardiothoracic surgeon, so I'll leave the heart surgery up to people who are. Hey, surgeons, why don't you let me handle your patients' insomnia? I imagine it will keep us both out of trouble.

Departing fact: As of the publishing of this book, no sleeping pill has ever been shown to increase daytime performance. On the other hand, the discontinuation of hypnotics has!

When Sleeping Pills Are Cool

While the vast majority of people who struggle to sleep do not need sleeping pills, they can be a useful tool in some situations. Understanding when a pill is useful and appropriate and when it is not is essential for the drug's effectiveness.

Sleeping pills are best used when people have specific and transient problems with their sleep. Here are some examples:

"I travel for work a couple times a month, and I really have problems sleeping in the hotels my company puts me up in. Outside of that, I'm fine."

"My husband was just diagnosed with cancer, and I am really struggling to relax and sleep at night."

"I just returned from two weeks in India, and I am struggling with jet lag."

It is okay to have trouble sleeping every now and then. In fact, it is more normal to have episodic difficulty sleeping than to live your whole life never having an issue falling asleep. We all have events in our lives that precipitate insomnia. That's okay!

Sleeping pills can provide a temporary solution to the problem of falling asleep (although be forewarned that some sleep pills worsen sleep quality once you do fall asleep). Think of them like a nasal spray. We all get congested occasionally. Using an over-the-counter nasal spray can be a perfectly appropriate way to get your sinuses back on track. But if you use them too long, your stuffiness can become a chronic problem. Sleeping pills are no different.

Occasional use for defined situations is perfectly appropriate. Using them every day . . . not so much. Remember, insomnia, like a stuffy nose, is a symptom, not a diagnosis.

The key with sleeping pill use is having a plan. What is the plan for your sleeping pill use? Is this a pill you will use for the next month while you deal with the loss of your family dog? Is this a pill you will take as you transition from day-shift to night-shift work and are forced to sleep during the day leading up to your night shift? Is this a pill you will take when you fly to China and have to sleep in a noisy hotel in Beijing? Regardless of why you are using the pill, you and your doctor need a plan.

An essential component of the plan is when you will not take the pill or stop taking it entirely. You may give yourself a month with the pill while you grieve. You may allow yourself to use the pill for a few days after your work shift changes from days to nights. The pill may stay in your overnight bag, being used only for travel. In all of these cases, the plan is dictating when you use the pill, for how long, and when you don't. That's a smart way to use sleeping medication.

But doctors everywhere have a big problem with this part of the plan. For many people, the plan seems to be "take one pill by mouth every night before bed until you see a bright light and deceased friends and relatives beckoning you to join them. Fill 30 pills. Refills: 600." That's how the prescription is written and it's a big issue. In other words, the plan seems to be to prescribe this pill for the rest of the patient's life, and this is not a proper plan for a sleep aid.

The problem is that doctors often don't discuss a long-term plan when they prescribe sleeping pills. It is strangely not missing from other problems doctors treat. Can you imagine seeing your primary care doctor for a nosebleed and the response to the problem being a cotton ball placed into your nostril with follow-up in a few days? This might be reasonable, but what would happen if you returned and the cotton ball was removed, creating a continued gush of blood from your nose? Would you be surprised if his action was exactly the same: more cotton and follow-up? How many visits would you endure before you would finally say, "Aren't you going to do something about finding out where this blood is coming from and how we can stop it?" The repetitive administration of sleeping pills over months and years without trying to figure out why there is a problem with sleeping is no different from more and more cotton!

  SLEEPING PILL SCAVENGER HUNT

1. Get out a piece of paper and a pencil.

2. Make a list of every sleeping pill you are currently taking. For this exercise, any pill you take that is specifically for sleep counts, even if it is not technically a sleeping pill (for example, Seroquel is an antipsychotic, but many people are being prescribed it for sleep). Give yourself a point for every one of them. If the drug is a controlled substance or you need a written prescription for it every time it is filled, you get 2 points.

3. Add to your list any drugs you've taken in the past for sleep. If you discontinued a drug because it was ineffective, you get a point. If you discontinued a drug because your physician was concerned about how much you seemed to need to fall asleep, you get 2 points. Again, any controlled substance is worth 2 points.

4. Add to this list the date you started and, when appropriate, the date you stopped each of these pills. For any drug you started more than ten years ago, give yourself 1 bonus point. For any drug you have taken continually for more than five years, give yourself 1 bonus point.

5. If you had issues discontinuing any medication (withdrawal), give yourself 3 points.

Congratulations. You now have a wonderfully complete list of the medications you have taken for your sleep problems. This will come in handy if you ever see a sleep specialist in your future.

In 2015, the San Francisco 49ers football team scored a league low 14.9 points per game on average. Did you outscore this team? If you didn't, did you come close . . . like within a field goal? Ask yourself one simple question, "Why aren't these pills working?" If you did, about that sleep specialist referral . . .

For many people who chronically use sleeping pills, the chemical effect of the pill is nothing compared to the psychological comfort it provides. In other words, the pill becomes their baby blanket. All three of my kids had a blanket (we called them "Boo"). Sleep was always great if said kid went to bed with Boo. Without Boo . . . take cover! I remember taking trips and while getting the kids ready to sleep in the hotel asking my wife where she packed the Boos and her response being "I thought you were packing them when you put the pillows in the car." This was typically followed by angry but silently terrified glances back and forth as the saucer-eyed kids started preparing their brains to be awake for the foreseeable future.

How can a ratty gray piece of material make such a difference in a child's sleep? Belief, habit, and fear! The use of a sleeping pill is usually no different. The patient *believes* it is useful and is in the *habit* of using it every night. Most important, the patient *fears* what will happen if she doesn't have her crutch. But if you've read this whole book, you know there actually is nothing to fear—except maybe taking the pill!

Most people who get hooked on sleeping pills do so innocently. They start for a good reason but have no plan in place to let them know when it is time to stop . . . so they never do.

Who needs a sleeping pill? This list is fairly short:

1. For brief periods of acute stress secondary to clearly identifiable stressors or sleep disruptors: loss of a loved one, loss of a job, divorce, chronic pain, and so on.
2. Environmental issues: sleeping in a hotel, camping with your family, or other episodic events that temporarily put you in an environment where sleep is difficult.

3. Shift work disorder: when an individual who, as a result of her job, sleeps at nontraditional times and has sleep disruption and sleepiness as a result.
4. Jet lag: trying to sleep in a location where your internal body clock differs from the external environment.
5. There are some who would add "primary insomnia" to this list. These are the people "who just can't sleep without medication." I believe there are patients who have higher-than-normal arousal than others, but to give a pill out of fear that a patient is not sleeping . . . I'm not buying it. Their sleepiness score is always less than mine! These patients need cognitive behavioral therapy, not a pill.

By looking at points 3 and 4, it is very clear that the timing of one's sleep can be a major factor in both the development of a sleep problem and the treatment of the problem. The next chapter delves into the incredible importance of schedules and circadian factors when it comes to sleep. In other words, understanding sleep, and knowing that you do, is an important first step. Getting yourself properly prepared to sleep by cleaning up your bad habits and ditching the pills is essential as well. Now the question becomes when should you sleep? Fortunately, I know that too!

CHAPTER 11 REVIEW

1. Sleeping pills are appropriate when they are used in specific situations for a specific purpose. Among these are acutely stressful situations, jet lag, and shift work difficulties.
2. But if you take a sleeping pill, be sure you have a plan for its use. Before you even start taking it, work out with your doctor under what circumstances you will take the pill, for how long, and when you will move on to some other form of therapy or intervention.

3. If you are currently hooked on sleeping pills, talk to your doctor about how to get off them. If you do, you'll sleep better.

Hopefully, you now have some new concerns about your sleeping pills. You've tried them all and they don't work. What can you do to improve your sleep? Let's think about cognitive behavioral therapy again. You are smart, you understand your feelings about sleep, and you use your bed just for sleep. Sleep restriction and schedule consideration are essential. Open your calendar app; we need to make some changes to your schedule.

12

SLEEP SCHEDULES

I'd Love to Stay and Chat,
but I'm Late for Bed

HAVE BEEN ASKED FREQUENTLY OVER the years the following question: "What is the single most important piece of advice for achieving your best sleep?" To me, it's easy: pick a wake-up time and stick with it!

When I ask you, "What time do you wake up in the morning?" the answer should be one simple time. If your answer to that question is, "I get up at 6:45 and usually go to the gym or run outside," you get a gold star.

However, you probably have a problem if you answer something like this: "I usually go to bed at 11:00 except on the weekends when I go out with my friends and we stay out until 2:00 or 3:00 easily. I'm usually up around noon . . . no later than 2:00 P.M. on those days. On Tuesdays I try to go to bed early, like 9:00 because on Wednesday, I have to get up early for this boot camp exercise class. On those days, I go out to my car at lunch and sleep for forty-five minutes. I feel pretty beat by the end of the week and often fall asleep early in the evening. If I do, I struggle to stay asleep and often have trouble falling asleep later. It's really hard to get up for work on Monday . . . I'm often late. Every now and then, no more than once a month, I'll use a sick day and stay home from work so I can sleep all day . . ."

184 THE SLEEP SOLUTION

Wow, I blacked out there a little bit from acute boredom during that long explanation. But I have to tell you, that was a real patient story. Optimally, an individual should have a consistent bedtime and perhaps more important a consistent wake time. Unfortunately, this isn't usually the case with individuals who have sleep problems. Sleep times can vary wildly in these people and they sadly do not seem to recognize that this haphazard lifestyle is a big part of their problem. In fact, they strangely often see it as working toward a solution.

Some people are in complete control of their sleep schedule. No matter what happens in their lives, they are up at 6:00 A.M. and soon after in the gym getting their BodyPump class on. These individuals are dogs who wag their tails (the dog is in control, and it wags its tail). Other individuals get up and exercise if their evening goes perfectly to plan, but if their sleep is problematic one night, their schedule goes down the tubes. If it takes them an hour or two longer than normal to fall asleep, they ditch their exercise plans and sleep in. For these individuals, their wake time depends on their sleep quality. They are not in control, so instead of the dog wagging its tail, for these individuals, the tail wags the dog. I call them "dog waggers." Their sleep schedule is dictated by their sleep successes or failures.

Here are some examples of dog waggers:

> "I went to bed early last night because I had a rough night of sleep at my girlfriend's apartment the night before."

> "My alarm went off at 6:00, but since I couldn't fall asleep until 3:00 A.M., I hit the snooze and called in to work later saying I was sick."

> "My wife was driving me crazy because I masked the basement last month but haven't painted yet, so I just did it . . . stayed up most of last night to do it. I took a huge nap when I got home from work, so now I'm wired."

When an individual sleeps in this way, all kinds of bad things are happening. You are teaching your body to sleep only when it's exhausted. Like a cow grazing for food, you're grazing for sleep. If you are independently wealthy and have no need to work, congratulations! Maybe the schedule of the world doesn't apply to you and you can keep up your freewheeling schedule as long as you wish. For the rest of us, the world is full of appointments, deadlines and plenty of times we need to be awake.

I joke with my patients all the time that if I can't fix their sleep problems, they should enlist in the military. The army is such a wonderful environment for sleep. They do everything sleep related perfectly. Their wake-up time is precise. Fall out at 5:00 A.M. Tired? You'll get over that quickly as you and your platoon head out for physical training. Change for breakfast at exactly the same time every day. Activities, more exercise, lunch, dinner, and finally back to bed so you can get up and repeat the exact same schedule the next day. Within a few days of boot camp, you will have all sorts of wonderful and exciting problems . . . falling asleep at night will most likely not be one of them. I always think about these soldiers when a patient tells me he has trouble settling at night because his "mind won't shut off." I imagine after a day of engaging in grueling exercise, being screamed at and belittled, and missing your family as you wonder what in the hell you've gotten yourself into, you might have some thoughts racing through your mind. But still, these men and women sleep.

So when should you go to bed? When should you wake up? I think you see by now why we have to start with the latter. You need to select a wake time that works for your life. If you have to be at work by 9:00 and your commute is thirty minutes, getting up an hour earlier might work for you. Unless you want breakfast or maybe a shower? Unless you want to exercise or maybe you have kids who need to get up for school? The point is to pick a time that's realistic. And be sure to include time to actually feel awake.[72] Nobody opens her eyes and feels like Mary Sunshine in the morning, at least nobody over three. So be sure to give yourself some time to go from groggy to human.

---

72 It's amazing to me how many people judge the quality of their sleep by how they feel immediately upon awakening. I personally feel like Han Solo after he's been released from that block of carbonite. But a few minutes later, I feel like myself.

One more important detail: There is no such thing as a good or bad wake time. Yeah, depending on whether you are a night owl or a morning lark, one schedule might work better than another. Are you a morning person? Getting up at 6:00 might be better than noon. Always been a real night owl? That 5:30 A.M. wake time to meet some friends for a bike ride might not be ideal. Brain schedule aside, I'm not here to judge. People in the South talk about sleeping in like they talk about sex: in embarrassed whispers. There is nothing wrong with being a night owl. It's not a sin.

Establishing a consistent wake time is the most important first step in setting your schedule and solving your sleep issues. Once you've picked a wake-up time, the big question is "How much time do I need to sleep?"

Did you know the average individual eats seven Chips Ahoy! cookies upon opening the bag? Okay, fine, I made that number up. But let's go with it. Now let's imagine we went to the mall and selected 100 random "average" individuals and gave them each seven cookies. Does that mean every individual would clean his plate? No. Some would eat fewer cookies while others would demand more. Should those who eat less worry about it? No.

We all need different amounts of sleep. Don't get hung up on magazine articles that demand you sleep eight to nine hours every night for your optimal health. There is a good chance that whatever number they are throwing out at you is not ideal for you specifically.

  ICE BUCKET SLEEP CHALLENGE

If you have difficulty falling asleep or staying asleep, this is a wonderful exercise you will thoroughly enjoy!

1. Determine what time you need to wake up and set multiple alarms for that time.
2. Fill a bucket with ice water and set it near your bed. Instruct your spouse to pour it on you if you don't wake up with the alarms scattered throughout the room.

3. Work backward from your wake-up time five and a half hours. This is your new bedtime. In other words, if your alarm is set for 6:30 A.M., your bedtime is now 1:00 A.M.! Exciting . . . you are going to get so much done!

4. The rules are simple. You can go to bed at your bedtime or any time *after*. Not sleepy at 1:00? Feel free to stay up as late as you like!

5. You must be up and out of bed by 6:30 or earlier. There is absolutely no sleeping in. Remember the bucket!

6. Napping is not allowed. Neither are sleeping at your desk, falling asleep before dinner, nor napping on the couch in the evening. Sleep is not allowed at any time other than your nightly sleep period: 1:00 to 6:30 A.M.!

This exercise is a tough one to complete. Conversely it is an easy one to give up on. Why would anyone adopt this schedule? How on earth can an exercise like this be the key to your sleep success? Hang in there . . . A seed doesn't sprout immediately.

Notice how little changes about your sleep for the first couple of days. In fact, this isn't really working at all, Doc. The only thing that seems to be happening is that I am really getting sleepy during the day and finding it harder and harder to stay awake until 1:00!

Exactly![73]

Chances are several things are going to happen to you when you embark on the Ice Bucket Sleep Challenge. The first is that it won't work at first. Our brains have a structure called the *suprachiasmatic nucleus*. This structure is the internal timekeeper of our brain. It functions to help us time virtually everything our bodies do. It regulates when we get sleepy and when we feel awake. It regulates when our bodies release certain enzymes and hormones. It regulates the fluctuations of our body temperature. And so on . . . These rhythms can take time to change, so don't be discouraged if your problem isn't solved on day one.

As time passes with this exercise, the sleep disruptions that were

---

73 Again, *exactly*!!

once there slowly start to fade as the body desperately tries to satisfy its sleep need by making the five and a half hours of sleep opportunity each night as efficient as possible. In other words, over time, provided you aren't cheating (and hitting the snooze ten times is very much cheating—stop using it), your brain will begin to exhibit a stronger and stronger drive to sleep upon going to bed because it is coming to the conclusion that if it wants sleep, those precious five and a half hours are the only time to get it.

Got a kid who won't eat his dinner? Take away all of his snacks and cut his lunch in half. Watch what happens over the next two weeks with dinner. The principle with the Ice Bucket Sleep Challenge is the same.

As time passes, the brain will start to adjust. Sleep becomes more continuous and deeper; a natural way to compensate for lack of sleep quantity is to increase sleep quality. Eventually difficulties getting to sleep or staying asleep are a thing of the past. The individual's biggest problem now is staying awake during the day!

Something else is happening. One of the most critical pieces of the puzzle is falling into place. Where there was once fear of not sleeping, there is now a growing confidence in the ability to get into bed and fall asleep. No pills . . . no iPod relaxation apps, no concoctions of valerian and melatonin . . . just getting into bed and falling asleep. As the consecutive nights of successful sleep initiation start to rack up, the individual continues to be anxious about her job, her loved ones, her favorite sports teams, Kim and Kanye's marriage—but she is slowly letting go of her anxiety about whether or not she will sleep.

The technique I just explained is called *sleep restriction*, and it is an integral part of cognitive behavioral therapy (CBT-I). Patients are often stunned when I tell them that to fix their sleep they may need to temporarily spend less time in bed. Some patients walk out of my office muttering about the sleeping pills they thought they would be prescribed. You know by now that sleep is a primary drive and no pills are needed. And those who endure the short-term pain will enjoy the long-term gain of a restful night's sleep—and learn a lot about how much sleep they actually need to feel good in the process.

That truly is touching, but as that problem fades into oblivion, their new problem of increased daytime sleepiness is becoming a real bother. No worries . . . we knew this would happen with the Ice Bucket Sleep Challenge. Here's the solution. Keep that wake-up time anchored at 6:30, but move the bedtime to 12:45 A.M. instead of 1:00. That will give the individual fifteen minutes of extra sleep every night or almost two extra hours every week!

What happens? Well, if that solves the problem of excessive sleepiness during the day, you're done! You officially seem to need five hours and forty-five minutes of sleep at this point in your life. Unlikely, but possible. If excessive daytime sleepiness hasn't gone away, more manipulations of your bedtime (not wake time) need to happen until you can get into bed, fall asleep in fifteen minutes or so, stay asleep, and feel good the next day. While there are very rare individuals who can survive on less than six hours oı sleep, usually patients who undergo the sleep restriction training tend to need six and a half to seven hours. Just remember, everyone has a different sleep need and that need changes (usually shrinks) over time.

Our body's circadian rhythm can be the key to constructing a schedule that can finally fix your sleep issues for good. While a well-constructed schedule can offer many individuals sleep salvation, a poorly constructed schedule can be the gateway to purgatory!

Circadian Rhythm Disorders

Sleep schedules and our circadian rhythm are important. When everything works well and is timed well, our bodies work as seamlessly as a beautiful symphony, each organ system coming into the song at just the right moment.

Imagine the symphony now without a conductor . . . or maybe a better analogy would be with a drunk conductor. The brass is coming in too early; the percussion is way behind. This is the mental picture I want to paint for you when we think about circadian rhythm disorders. The body is too early, it's too late, or it just does not seem to have any clue.

Shift Workers: The All-Star Team of Screwed-Up Sleepers

If I could put an asterisk next to any section of this book, it would be this one. Nobody faces more challenging sleep issues than shift workers.

Shift workers are, by definition, any group of individuals who work "nontraditional" hours, meaning any job that doesn't happen between 9:00 A.M. and 4:00 P.M. In other words, you don't need to work the graveyard shift to be a shift worker. You can work from 2:00 P.M. to 11:00 P.M., or work a normal shift some days and a nontraditional shift others. The possibilities for sleep disturbance are endless.

Millions of people participate in shift work. In America, approximately 15 percent of workers do not work traditional hours. Many of these people handle the work gracefully, but about a quarter of these workers struggle with the unnatural schedule. Many of these individuals have problems as a result. Heart disease, mood problems, weight issues and cancer have all been linked to shift work.

It's not a fun gig. And those are just the health consequences. The consequences on home life can be equally devastating. Trying to coordinate schedules and having the rest of your family understand them can be very taxing on individuals. I'm going out on a limb here when I say that in particular for women who are shift workers, it is especially hard because in addition to their jobs they are still responsible for shopping, cooking, housekeeping, and child care in many situations.

As a rule, the older we get, the more difficult shift work becomes because as we age, we become less night oriented and more day oriented. This is important because night owls deal with changing schedules better than morning types, so as we become more morning oriented, we become less suited for shift work.

Workers who are forced to sleep during the day get less sleep than normal-schedule workers, are sick more, and have greater challenges in their personal lives. The 9:00-to-5:00 world is often "not open" when they are awake. For them to get to the bank or participate in their exercise class, they often are forced to sacrifice their sleep. This means that shift workers are frequently moving back and forth between schedules. Remember that suprachiasmatic nucleus? Without a schedule, it has a hard time figuring out the body's timing and thus shift workers are often sleepy when they should be awake and awake when they are trying to sleep.

It is essential that shift workers' sleep—when they can get some—be normal. Without power over the sun, it is up to the worker to re-create the darkness of night to get some sleep, even if he has to do so when the sun is rising in the morning. Your brain isn't stupid . . . It knows what's up. Care has to be taken to eliminate light.

The use of light exposure and avoidance of light can be helpful for this population. Their schedule also becomes critical. It's fairly easy to go to bed at night when it is dark, all of your friends and family are asleep, and television programming is basically home shopping and syndicated reruns of *Friends*. There's nothing to do. Contrast that with the individual who has finished up her night shift at 7:00 A.M. and is headed home from the nuclear power plant. The sun's coming up. It's a beautiful day. She's passing the gym. She needs to do some grocery shopping because she's out of milk, OJ, and eggs. Once she gets home, there is a militia uprising being televised on CNN as her kids fight over who gets the last Oreos for their lunch. Her partner has an early meeting (as usual) and can't drop the kid off at school. "Can you drop her off real quick before you lie down?" Bills, bills, bills . . . Get my drift. Sleeping during the day is hard.

I'm not a fan of sleeping pills, but this is a group for whom they can be quite helpful. Shift workers can often benefit from medications not only to help them maintain wakefulness but also to initiate sleep. The diagnosis of shift work disorder is currently an FDA-accepted diagnosis and a valid reason for prescribing medications like modafinil for wakefulness promotion. Should shift work disorder be a medical diagnosis on the same level as gout or ringworm? That's not for me to decide. What I will say is, this disorder poses more of a threat to a patient's life when he gets off work and behind the wheel of his PT Cruiser than do gout and ringworm combined. Medications designed to improve wakefulness can not only improve work productivity but potentially help ensure that driver makes it home alive.

Shift work is costly to employers because it is hard on workers. Shift workers lose out on more sleep than their day-working counterparts. It is hazardous to worker health in many ways, and in my opinion, with the amount of research coming out about the dangers of participating in it, it won't exist in the way we know it twenty years from now. The worst offenders have already started to slowly reform. When I started my medical residency, there were zero hour restrictions on how much we could work. Know why it's called *residency*? It's because doctors a generation before me actually resided in little apartments within the hospital. The hospital was their home. During the final year of my residency, the powers that be put an end to the unrestricted work and limited residents to an eighty-hour workweek. Change is slow.

To me, shift work is like asbestos. Asbestos is a kick-ass insulator, and the fiber is also resistant to fire and performs well as a sound absorber. It has all kinds of helpful chemical properties and it is abundant. The problem is that it will also kill you.

Shift work is similarly a kick-ass way to increase the productivity of your business over a full twenty-four-hour workday. Workers are abundant and generally require less management to be on staff at night. In fact, as you know now from reading Chapter 1, both asbestos and shift work can lead to death. I'm sure it was quite disheartening to many when it was determined that a very common material used to insulate homes and buildings everywhere caused cancer. I can imagine that there was an overwhelming feeling of "How in the world are we going to fix this massive problem we've created?" When it comes to shift work and the discoveries of its detrimental effects on sleep, I feel the same way. How can we fix a problem that is so central to the way our culture is structured?

Delayed Sleep Phase Versus Advanced Sleep Phase

I've always been a night owl. I love staying up late. I frequently do my best thinking late at night. I felt drawn to a late schedule even when I was young. It's probably why I'm a doctor. I'm no genius, but it was always easy for me to stay up late and be pretty functional while the geniuses were struggling to keep it together.

As we touched on earlier in this chapter, we all have our sleep timing preferences, also known as our chronotype. *Chronotype*, loosely translated as "time type," is just that: It's the type of time our brains prefer.

When you think about an individual's sleep, there are two main variables to consider. The first is how much sleep an individual needs. We covered that way back in Chapter 2. The second variable is when we want to get that sleep. That is what a chronotype represents.

Before you start stressing about what chronotype you want to be, guess what. It's kinda already been decided for you. Your chronotype is influenced by genetics. Specific genes called clock genes influence our chronotype. Age seems to be a factor as well, with individuals tending to be night owls in their younger years and becoming progressively more morning oriented with advanced age. Chronotypes are not usually absolute and can be manipulated, at least transiently, with disciplined changes in light exposure, meal timing, exercise schedules, social interaction, and sleep schedules.

The technical term for a night owl tendency is being *phase delayed*. A morning lark is someone who is *phase advanced*. Being phase delayed or phase advanced does not represent a sleep disorder necessarily. However, when an individual's chronotype is unable to adjust to meet the needs of his or her work or school schedule, then it can be diagnosed as a circadian rhythm disorder.

Many young people are phase delayed. They like to stay up late chatting in code with each other about their lame parents, calling each other BFF, and Snapchatting. Finally, at 3:00 A.M., they fall asleep. Unfortunately for them, school starts early—really early in some places. This has created a situation in which first period starts to look like a scene from *The Walking Dead*, with sleepy students everywhere—unmotivated, lazy, disengaged. Just as the bell rings to dismiss them for the day, the kids start to wake up.

In this example, some of those kids might really struggle to excel in those circumstances. I certainly never felt particularly great during early classes in college or medical school. These individuals can be classified as having a circadian rhythm disorder, specifically delayed phase type.

Advanced sleep phase disorder is different. This is about your dear grandmommy in Sarasota, Florida. She has a great day on Lido Beach, gets home for some lunch, watches the BBC *Newshour*, and then she's ready for bed. What? It's not even 8:00 P.M. Okay. Well, good night, Grandmommy. The next thing you know, it's 4:00 A.M., and she's in the kitchen with the Vitamix grinding through ice, kale, and coconut milk for her breakfast. Why is she up so early?

Grandmommy is up at 4:00 because her circadian rhythm is very advanced. So advanced that she feels frustrated sometimes that she awakens between 2:00 and 3:00 in the morning and cannot get back to sleep. The combination of her age reducing her need for sleep, her lowered activity level, her permissive schedule that allows her to nap during the day, and her advanced chronotype leading to an earlier bedtime and wake time can lead to the disorder.

Non-Twenty-Four Disorder

Because light is such a strong regulator of the circadian system, individuals who are blind have a hard time maintaining a proper circadian rhythm. Without light to set or *entrain* their circadian rhythm, their sleep and other circadian processes (like eating) can get out of whack, leading them to have trouble with their sleep. There are specific drugs for this population that can be very helpful.

Treating Circadian Rhythm Disorders

For all of these circadian disorders (shift work, delayed sleep phase, advanced sleep phase, and non-twenty-four disorder), medications to help with sleep can be helpful. But again, if you plan to use them, remember you first need to have a plan in place for doing so.

While medications to help with sleep can often jump-start the treatment of circadian rhythm disorders, they need to be coupled with other therapies for long-term success. The most important therapies for everyone except non-twenty-four patients involve light. Because light promotes wakefulness, Grandmommy needs it later in the day to help keep her awake past *Jeopardy!* Her grandson needs it when he first wakes up and even while in his earliest school periods to keep him alert. For shift workers, medications and light are often used in conjunction to help promote wakefulness. These interventions can be incredibly important for helping individuals stay focused during work or while driving.

 PRODUCT SUGGESTION

WITH THE ADVENT OF CHEAP and bright LED lighting, light boxes for the treatment of circadian disorders have become much more affordable and accessible. I love the Lightphoria by Sphere Gadgets. It's available online and is brighter than the sun! If you don't want to be stuck in one place, consider the Re-Timer, a rechargeable and wearable blue-green light that makes you look like Tron.

And don't forget those Uvex glasses I mentioned earlier. They make sure the light from your computer does not affect your sleep by blocking the blue light that keeps us awake. Blue light–blocker sunglasses can get you prepped for sleeping after your night shift.

Jet lag is a form of circadian rhythm disorder as well, so make sure earplugs and a comfortable eye mask travel with you everywhere. Instead of a bulky neck pillow to support your head during your next flight, consider a NapAnywhere (www.napanywhere.net). This collapsible head support is comfortable and lightweight, and it folds up into the shape of a Frisbee to easily fit into a thin computer bag. I go nowhere without mine!

Exercise, intelligent, sleep-friendly dietary selections (mentioned in the last chapter) and adhering to a proper schedule are all integral in terms of keeping circadian disorders from creating problems in your life.

....

CHAPTER 12 REVIEW

1. The single most important thing to achieving successful sleep is to have a consistent wake time.
2. Once you have that, you can choose the going-to-bed time that works for you.
3. One way or another, everyone, young or old, needs a schedule to sleep by.
4. This schedule does not have to be eight to nine hours for everyone.
5. Shift workers need especially sound schedules, and even then they may require further help with their sleepiness and wakefulness.

Sometimes, even the best schedules can get off track and some additional sleep is needed. How should we nap? Where should we nap? When should we nap? Answers lie ahead . . .

13

NAPPING
Best Friend or Worst Enemy?

LOVE A GOOD NAP. FEW people don't. Somehow the business of sleep feels like work to many people at night, but the little bonus snooze that happens when you stretch out on the couch Saturday afternoon after taking the kids to soccer practice in the morning feels different . . . more relaxed . . . more decadent.

Outside of the question, How much should I sleep? there is no question I get asked more often than, Is it okay to nap and for how long? Great question. Let's talk about it.

To figure out the role napping should play in our lives, we need to come to some conclusions about our sleep, and what makes it "good." One of the biggest factors that people relate to good sleep is efficiency. Sleep efficiency is basically a math equation:

Time Spent Asleep ÷ Time Spent in Bed x 100 = Sleep Efficiency (%)

Admit it—that's really easy. It's just the percentage of the time you are sleeping when you are in bed. And what is a normal sleep efficiency? There are minor disagreements here, but for our purposes, we will say 85 to 90 percent is the target. Why not make 100 percent the target? To answer that, let's look at some simple examples.

Consider a person who goes to bed at 9:00 P.M. It takes her 1 hour to fall asleep. Once asleep she will usually not awaken for about three hours. At that point she hits the bathroom, checks her email, and goes back to sleep about 30 minutes after she originally awakened. She will then sleep pretty soundly until 7:00, at which time she will awaken, sit there for about 45 minutes, and then get up. Calculators ready!

So this person is going to bed at 9:00 P.M. and getting up at 7:00 A.M. This gives her a time in bed (TIB) of ten hours. Her sleep time is considerably less:

10 hours − (1 hour + 30 minutes + 45 minutes) =
7 hours 45 minutes

So to plug into our sleep efficiency equation:

7.75 hours ÷ 10 hours × 100 = 77.5 percent efficiency[74]

In this example, despite the sleeper getting almost eight hours of sleep, her sleep efficiency is relatively low. For this reason, she will typically feel pretty bad in the morning. When an individual comes to my office complaining of "not sleeping," more times than not, what they are really complaining of is poor sleep efficiency. Does poor sleep efficiency always equal sleepiness? Not at all. In 2000, Kenneth Lichstein found in his extensive review of insomnia literature a consistent lack of daytime impairment or sleepiness. Patients who complain of not sleeping or poor sleep efficiency routinely have Epworth Sleepiness Scale scores that are perfectly normal. Despite this, do these patients "feel" poorly? Absolutely. Spending twelve hours in bed to get seven hours of sleep may not leave you feeling sleepy the next day, but it is often going to leave you feeling as if you'd been hit by a train.

When I was in residency, we would spend all day and night on call. If call was light, we slept. If call was busy, we got sleep only here and

---

74 Classically, sleep efficiency is calculated by using a TIB that begins once a patient falls asleep. For this exercise, I am purposefully using the time it takes to fall asleep to better illustrate the point.

there throughout the night. I was often struck by how poorly I felt after nights during which I got five or six broken-up hours of sleep. Like this example, the amount was okay, but the efficiency was awful. Remember, our target is 85 percent, so 77.5 percent feels pretty rotten.[75]

An interesting thing to note about this example is the fact that because this person feels rotten, she will often come to the conclusion that she is sleeping poorly. While this is an appropriate conclusion, her solutions for fixing the problem can be far from appropriate. These fixes are usually one of two things:

1. "I'm so tired, so instead of going to bed at 9:00, I'm going to bed at 8:30 to get some extra sleep."
2. "My sleep was terrible last night. I'm taking a nap."

The going-to-bed-early fix is something I see on a weekly basis. It uses the same logic that if you are not hungry for dinner at 7:00, it probably makes sense to go to the restaurant an hour early, so you can get a little extra food.

The napping does make sense because the individual is often quite tired after sleeping so poorly. However, what will the nap do to the efficiency of the upcoming night?

This chapter is about napping, not math and early dinner reservations, so let's revisit the original question: When should we nap and for how long? The answer: It's okay to nap when:

1. Your nighttime sleep is efficient yet despite this, you still feel sleepy (not fatigued—sleepy).
2. And the napping does not disturb your schedule for the upcoming night.

So what do we think about the example of the woman given

---

75 A sleep specialist once told me that sleep is like a symphony. Imagine going to a concert where the orchestra stopped playing every twenty minutes, even if they were in the middle of a piece. Imagine them taking these breaks over and over during the performance. You'd be very frustrated by the end of the evening, even though they technically played every note they promised they would play. Sleep, like music, is most powerful when it flows uninterrupted.

earlier? Is this an efficient night of sleep? Sorry, but 77.5 percent does not meet the criteria, so in this case, napping is not appropriate, even though this person probably really wants to take a nap. I know you're thinking it's pretty mean of me not to allow this poor soul to "catch up" a little bit on the sleep she missed out on last night, but if we look at it in a different way, you'll see it's the only rational thing to do.

I love to compare sleeping to eating (as I'm sure you've noticed). When it comes to napping, there is no better analogy. Imagine you have a kid in your house who is a "bad eater." He picks around at his dinner and fusses about not being hungry, completely impervious to the "starving kids in [insert third-world country]" speech. Night after night, the dinners become an all-out struggle to get some nutrition in his gut. In a state of total helplessness, you call the doctor, desperate for advice. As the two of you talk about the kid's daily routine, you mention something that raises the doctor's eyebrow.

"So he gets off the school bus around 3:30, has his pizza snack, and goes outside to play. A few hours later, he has dinner and that's when it all falls apart. I just don't get it." You sigh.

"What is his 'pizza snack'?" the doctor asks.

"Oh nothing. He just likes to have a few slices of pizza when he gets home. It's kind of his thing."

At this point the doctor would probably suggest that the kid's eating problem is not really his fault. Perhaps the big pizza meal is affecting his appetite a few hours later when he sits down to eat. In other words, to borrow from a previous section of this book, his need to satisfy that primary drive (hunger) is weak because he has already eaten.

Getting back to sleep. What are some reasons a person would exhibit a 77.5 percent sleep efficiency? One big reason is that the sleeper is "snacking" when it comes to sleep. What's another word for snack sleeping? Napping.

When an individual is struggling to maintain his sleep at night, the last thing we want to do is to expand the period in which he is trying to sleep (for example, by going to bed earlier or sleeping later). We also do not want to add in napping periods, as they will invariably reduce the sleeper's drive to sleep at night.

A nap is typically intended to complement or enhance an efficient night of sleep. It is not meant to make up for lost sleep when the sleeper had the opportunity to sleep but did not.

This is so important, I'm going to repeat it:

A nap is not meant to make up for lost sleep when the sleeper had the opportunity to sleep but did not.

Without a doubt, this is the biggest mistake people make in terms of their sleep, and frankly it's a killer when people retire. Why? Because there is nothing keeping the retirees from napping during the day when their sleep is poor at night. Their excessive naps lead to the inability to fall asleep at night when they want to, and the cycle invariably worsens.

Let's consider a different person. This guy goes to sleep at 12:30 and falls asleep immediately. He sleeps soundly until the alarm goes off at 6:00 A.M., and he's in the gym thirty minutes later. He's a Pilates animal for forty-five minutes, showers, and is in the office doing hedge-fund things all morning. By 11:30, he's tired and wants to take a fifteen-minute nap. How do you feel about this plan? Let's reexamine our nap criteria:

1. Does Mr. Goldman Sachs sleep efficiently? Hell yes, he does. The guy barely moves . . . total "princess sleep."[76]
2. Is this guy's fifteen-minute power nap going to affect his sleep later that night? Probably not. First of all, his nap is going to last only fifteen minutes. Thinking back to our bad-eater kid, it would be like swapping his plate full of pizza slices for a small handful of grapes. That's not spoiling anyone's dinner, and a fifteen-minute nap is probably not going to prevent this shark from sleeping when he hits the bed.

A word about sleep efficiency because this guy's efficiency is probably approaching 100 percent. Unlike everything else in this guy's

---

76 Princess sleep is that thing where you move so little when you sleep that you can literally make the bed with one little fold of the sheet.

life, when it comes to sleep efficiency, more is not necessarily better. An efficiency of 85 percent is great, and 90 percent is probably okay too. As sleep efficiency goes up from there, it's not especially a good thing. I know what you are thinking: a 100 percent on your AP European history test is way better than scoring a 90 percent. As Americans we shoot for 100 percent always! Why is an A+ not good for sleep efficiency? Well, for starters, humans wake up when they sleep. That's not only okay, but normal. Even if you are not aware you are waking up, you do, so aiming for 100 percent is an unrealistic goal. Also, consider what happens when you are really sleep deprived . . . like up all night for the past two days trying to get packed for a big trip or working on your taxes as the April 15 deadline approaches. What happens to sleep efficiency then? It gets really high. Is this a good thing? Does going to sleep at 4:00 A.M. and awakening at 6:30 A.M. and having virtually 100 percent efficiency mean your sleep is great? Not really, so beware of extremely high sleep efficiencies because they often just indicate sleep deprivation.

The other thing to consider in this case is nap timing. This guy's nap is happening before lunch. Even if the nap does its job of reducing his sleepiness, he has plenty of time to find some more sleepiness before he retires at night. Sleep doctors have a saying: An early nap adds to the previous night of sleep but a late nap subtracts from the upcoming night of sleep. I have never seen a study proving this, but it makes sense to me, and because it's my sleep book, we're going with it.

While a nap that occurs early in the day is best, to really make it effective, it is vitally important that the nap be scheduled. Remember, the brain prefers to anticipate something, not react to it. A nap is no different, which is why a scheduled nap always works better over the long haul when compared to a random nap.

If you think about it, this makes sense. Remember how important it is to have a consistent wake time? A nap should be no different. Pick a time for it to end that is the same every day. This does not mean you have to nap every day; it just means when you do, it should be at the same time every day.[77]

---

77 And parents, if you are struggling to get your little kids to nap, this is the first thing to consider. Are your kids napping on a consistent schedule, or simply "whenever they'll go down"?

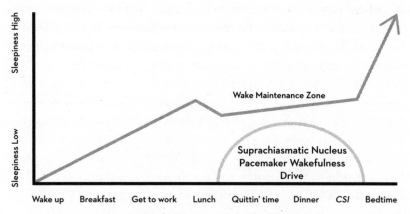

Figure 13.1. Exploring afternoon nap timing.

One more quick flashback. Do you remember the picture shown above?

If you look closely, there is a little sleepiness mountain peak shortly after lunchtime. This is the time when we experience a natural increase in our daytime sleepiness. Many sleep researchers and many cultures believe this is not only a great time to nap, but that we were evolutionarily designed to nap at this time.

The duration of your nap is important. A twenty- to thirty-minute nap is an ideal duration to provide a boost to your wakefulness without leading to post-nap funk (PNF). PNF is the fuzzy, dull, slightly headachy feeling one gets after a nap goes a bit too long. When an individual naps too long or naps in an erratic or un-scheduled way, the brain can enter into deep sleep. Awakening from this deep sleep feels awful, so basically PNF is your brain falling into deep sleep and not wanting to come out of it. Deep sleep is just that good.

Another blast from the past:

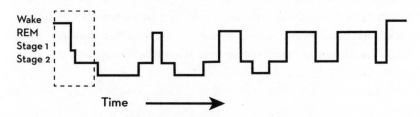

Figure 13.2. Return of the hypnogram.

Recognize this chart? Well, you should! It's from Chapter 4. Notice how sleep initially begins in lighter stages of sleep (in the dashed box). Ideally, a nap should encompass only the two lighter stages of sleep, like this.

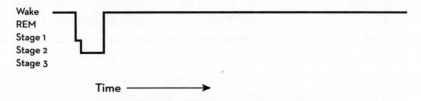

Figure 13.3. Hypnogram of a good nap.

Now look beyond the dashed box shown in Figure 13.2. What stage is coming up next? That's right, deep sleep. If the napper isn't careful, the refreshing pop of a quick jaunt through light sleep will be replaced by a mind-numbing descent into deep sleep. Now, the napper has woken up during N3. No wonder she has PNF!

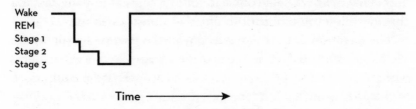

Figure 13.4. "Why did you let me nap so long?"

Like your awakening from a night's sleep, a nap should always have a definitive ending. In other words, the nap should always follow a schedule. "I nap every afternoon from 1:00 to 1:25." Regardless of how well you sleep, or don't sleep, the nap ends at 1:25 on the days you elect to do it. For extra pop, terminate your nap every day with some sunlight and some exercise. If sunlight is paired with a nap termination that is consistent, it creates a more powerful experience for your brain.

Setting the stage for a nap is essential. Nowhere is this more apparent than the napping I see going on within professional

sports. These athletes have the best of everything and the resources to have whatever they need to perform at the best of their
abilities. Their training facilities are showrooms of the most tech-
nologically advanced equipment, most nutritious and metaboli-
cally balanced food options, and most high-end comfort features.
Contrast that image with a full-grown man using a towel as a
makeshift pillow to sleep on the floor of a Major League Baseball
team's supply closet. This is a true story. I literally found him asleep
in a small pantry full of OhYeah! nutrition bars and canisters of
whey protein powder. "Get out!" he groaned as I opened the door
looking for a snack. I got out.

Most people would not sleep on the floor of a supply closet overnight, but when it comes to napping, people take what they can
get. This guerrilla napping is rampant.[78]

Why was the player in a supply closet? Two reasons: (1) Napping
requires a quiet, dark space, and frankly, this was the only place
he could find that fit those criteria. (2) Napping means you are
weak and lazy, so of course, it needs to be hidden from the public.
When you are choosing your nap setting, I hope only the first is a
consideration for you. If both of these things need to be considered, is there something you can do to make your higher-ups more
aware of sleepiness in the workplace and how much it affects your
overall productivity? There is a reason progressive corporations
have nap rooms!

When choosing a place to nap, a quiet, dark setting is preferable.
Find a place where you won't be disturbed. For me, when I nap, my
phone is off and my assistant, Tammy, knows that outside of my
wife, nobody gets through during that twenty-minute period.

I take my nap in a dark, quiet room. Although my office is very
quiet, I still use earplugs or some sort of sound machine to condition the noise level of my nap room.

---

78 Guerrilla napping is basically any kind of irregular napping in which you are using
the resources around you to nap, often in a clandestine fashion and often in a way that
is less complete or comfortable than the way you sleep at night.

PRODUCT SUGGESTION

FOR YEARS, DOHM HAS MADE the white noise machine I consider to be the industry standard. For a more varied array of sounds, I also like the Sound Oasis sound machine. It is small, portable, and battery operated, in addition to having an AC adapter. Besides white noise, it also has sounds of thunderstorms, the ocean, and woodland creeks as well as more popular sleep-inducing sounds (alpha chimes). It also has an earphone jack so others don't have to hear the birds chirping. And speaking of birds chirping, this device does a great job preventing looping of the sounds so your brain does not pick up on patterns. It truly seems natural and organic.

I have a chair that reclines in my office and has a hidden leg rest, so I can fully stretch out. This is important because it takes twice as long to fall asleep sitting up as it does lying down.

Get comfortable. Have a real pillow, not a folded-up athletic towel to rest your head on. If you are a lavender user at home, this is an excellent place to have lavender as well, as the smell will suggest to your brain that you are in fact home and sleeping in your bed.

Have a blanket available. I have one that feels like fur. It's fake fur, of course. The unusual texture of the blanket provides another clue to my brain, like the smell of lavender, that it is time to sleep. If you wear a lot of fur in your day-to-day life, this won't work as well.

 SLEEP PRODUCT

LAVENDER IS A TRIGGER FOR sleep if you use it consistently in your bedroom, but some small studies suggest lavender might be in and of itself sleep promoting. In a small, well-constructed study by George Lewith of the University of Southampton, lavender, when infused into a bedroom, appeared to help promote sleep. A 2014 study performed at Johns Hopkins Hospital found improved sleep in an intensive care unit population exposed to lavender. My favorite lavender spray is the two-ounce Aura Cacia Pillow Potion. I like this because I can keep it in my carry-on (for spraying in my Marriott Hotel room), and it won't get confiscated by the airline TSA agents.

Here's an additional tip. Need a good baby shower gift? Buy someone a stuffed animal filled with lavender. These little guys can be gently heated in a microwave before being put in to sleep with the baby. While everyone else is buying little onesies that look like sailor suits that he'll outgrow in three weeks, your gift will help him—and the parents—sleep. They will love you, and the smell will help mask the scent of the Diaper Genie!

As a note, the American Association of Pediatrics recommends that no stuffed animals be used in a crib until the baby is one year old or older.

Now it's time to sleep. My trick is to never approach a nap with the intention of sleeping. My goal is to lie flat in that dark room and think about whatever pops into my mind. I'm not racing from the crazy thoughts. I'm embracing them. Go ahead and think about the grocery list. Plan out a speech for your boss about why you deserve a raise. Many people struggle to sleep because they can't turn their mind off. Don't be that person. Just leave that mind on and running. Don't worry—if you need to sleep, you'll conk out for your nap. And even if you don't, you'll get up feeling rested.

Sleep Debt and How to Repay It

Sleep debt is a big topic in the sleep world and one that journalists love to write about in magazines. A sleep debt is exactly what it sounds like. It is a night in which you obtain an inadequate amount of sleep. In other words, you stay up too late reading a novel, you work two jobs, or your flight gets delayed and you have to overnight in Atlanta. Whatever the reason, you have slept inadequately.

Individuals collecting a sleep debt are quite common in today's twenty-four-hour culture. A French study of 1,004 twenty-five- to forty-five-year-old subjects estimated 38 percent of the subjects surveyed to be acquiring a sleep debt or be sleeping fewer than six and a half hours on a routine basis.

A chronic sleep debt is not good, and this should not be a surprise to you if you've read this book. Keep in mind that when we talk about sleep debt, we are not talking about insomnia. We are talking about someone who is purposefully depriving himself of sleep. (Okay, you didn't set out to binge-watch *House of Cards*, but it won't watch itself, right?) Recent studies show that this sleep debt has dire health consequences, including weight gain and impaired blood sugar control.

The question is, Once that sleep is lost, can I replace it? Does that nap adequately repay the sleep debt? If so, and here is the big question, how long do I have to pay that debt back? Do I need to nap the following day? Within the next week? Two weeks? A month?

The short answer is that we really don't know for sure, but evidence is pointing us to the conclusion that short-term sleep debt can be made up for if we do it relatively quickly. While according to a 2008 study, one night of makeup sleep may not be enough to counteract the ill effects of a modest sleep debt, a 2016 study by Josiane Broussard showed that two nights of makeup sleep (after four nights of sleeping four hours and thirty minutes) seemed to return insulin levels and diabetes risk to normal levels.

Here's my take, and it represents about 50 percent scientific certainty and 50 percent educated guess. I think we can repay modest sleep debts as long as they are repaid quickly and in full. Stay up late ringing in the New Year in Times Square? No problem, just make sure you repay that debt within the next few days, because as time passes, I think the window for undoing the tiny hit to your body vanishes. In other words, the window to repay all of the nights I spent awake on call as a medical resident is officially closed and whatever it did to my health is probably done. All we can do is look forward!

CHAPTER 13 REVIEW

1. Napping can be okay if done wisely. Just as you should have a plan for consistent wake time in the morning, have a plan for your nap wake time as well.
2. Napping is okay if it is efficient and satisfying. It's best if done early in the day and kept under thirty minutes.
3. If you stay up too late one night, make up your sleep debt as soon as you can.

You know it all now and you are controlling what you can control: your attitude, the amount of time you sleep, the timing of that sleep. You are a star. Now, let's shift our focus to some things outside of your control. Let's start with that noise coming from your bedroom at night that sounds like a cross between a chain saw and a *Walking Dead* zombie. . . .

14

SNORING AND APNEA
Not Just a Hideous Sound

INALLY WE GET TO THE meat and potatoes of sleep disorders.[79] This disorder is sleep apnea and his cheeky sidekick, snoring.

Snoring probably affects one third to one half of people over the age of thirty. When I turned thirty, I remember Ames telling me that if I rolled over on my back, I sawed some logs. For years, like most men who come to my clinic, I assumed my wife was lying. Everyone knows that women have nothing better to do than to accompany their husbands to doctors' appointments and make up stories about their spouses' breathing at night. (Again, this is sarcasm. You know she's telling the truth.)

My snoring for some time seemed to come and go. If I was up late studying in medical school, I tended to get more memos from the boss that I was snoring. At one point during a particularly stressful time, it got so bad I looked up treatments for snoring on the Internet, which was no small feat with my dial-up modem.

The first treatment I tried was the "sew-a-tennis-ball-onto-a-shirt" method. This method still exists and has been refined a bit,

---

79 Fittingly, meat and potatoes often play a role in the development of this sleep disorder.

but the premise stays the same: make it really uncomfortable for the snorer to roll onto his or her back. This method is effective for some because the airway is in a more stable position when one is on one's side.[80] This is helpful for those afflicted with *positional snoring*. What I found it most effective for was eliminating the comfortable feelings in my back I had grown accustomed to all of my life, as I would invariably awaken directly on top of the tennis ball. Somewhere between the term *kink* and that special finishing move in the Ultimate Fighter video game where one player rips the spinal column out of his opponent, there exists a description for the pain I felt after waking up on the ball.

Ames was amazed that I could maintain my sleep on that ball. Remember primary drives? When you are Medical School Brand sleep deprived, you can sleep through some serious discomfort: dental procedures, spinal taps, a lengthy ballet recital in which your daughter is literally onstage thirty-eight seconds. I had no difficulty rolling onto that ball and staying on top of it.

Never one to admit defeat, my plan B was to wear a backpack at night containing a basketball. I went with the red, white, and blue ABA basketball rather than the standard orange. It made me feel a little cooler, which was important, as there was little cool about the way I looked. My wife must have felt the same way because there was little Quasimodo love coming my way during this trial.

Did the backpack keep me off of my back? Of course it did. Did I feel like Luke Skywalker with Yoda constantly on my back telling me to feel the Force? Yeah, I felt a little of that too. It was hard to keep that thing on at night. I would often awaken in the morning, on my back with the L.L.Bean bag on the ground. I tried elaborately tied straps and knots, but like Houdini, I could not be contained. As if it were some kind of magic act, I would actually have Ames inspect my knots before my nightly performance. My magic was so good, the secrets of my escapes are actually still unknown to me.

Despite my failures, I was at least becoming more comfortable with personal physical restraints, so I thought it was time to move to more extreme measures. As a medical student, I had access to

---

80 If you are a genius infant reading this, ignore it and "get *Back* to *Sleep*." Sleeping on your back greatly reduces the incidence of SIDS.

all kinds of medical supplies and equipment. Among the drawers of rubber gloves, Surgilube, and cards for detecting blood in stool (the prestigious items that medical students get lots of experience with), there were disposable psychiatric restraints. Thank you very much—don't mind if I do. Hours later I was showing Ames.

At this point, I think she felt this was less bizarre than the backpack, so she just rolled her eyes, thoroughly bored now with plan C.

Preparing for bed that night there was anticipation in the air as I bound myself to the bed, facedown on my stomach. Ames graciously turned on my alarm since I couldn't. We kissed good night with minimal fuss and turned out the lights.

I lay there for a bit, not finding it nearly as uncomfortable as I expected I would. As the last traces of consciousness dripped out of my brain, Ames whispered in the dark, "What if the condo catches on fire?" Damn it.

I eventually went to sleep and slept well. I didn't David Blaine my way out of the restraint and fortunately did not awaken with the urge to urinate. (Note to self: Steal a handheld urinal from a hospital supply room today.) I felt no different, Ames was pleasantly surprised, and it seemed my snoring problems were solved. I repeated the process and eventually trained myself to sleep in one position, on my side.

Positional snoring is one thing, but obstructive sleep apnea syndrome is another. You can think about snoring as a loud sound associated with a vibrating airway. Apnea is when an airway is closing off. In other words, apnea affects breathing or how much oxygen a patient is getting during the night. Can you snore without having apnea? Absolutely. Can you have apnea without snoring? In some cases, yes, but typically snoring will be a tip-off as to the potential for apnea. In fact, the sudden disappearance of snoring might alert you that your bed partner is not breathing. The more severe an individual's breathing problem becomes, the less noise he may make.

Our brains use a lot of oxygen. Despite our brains weighing only about six pounds, they use a full 20 percent of our body's oxygen. If oxygen is oil, our brain is the United States—highly dependent.

Because of this dependence, our brain gets cranky when it is deprived. When someone has sleep apnea, she is repeatedly depriving her brain of oxygen throughout the night. In some cases, these spells of not breathing can happen twenty, forty, sixty times per hour or more. Much more in some cases.

So how does this relate to sleep? Simple. With each breathing disturbance, your brain has a decision to make. Stay asleep and let the suffocation continue or wake up and take a breath.

The effects sleep apnea has on sleep quality are equally problematic. Remember those graphs showing the various stages of sleep? Remember how deep sleep made us feel rested? When an individual is struggling to breathe and waking up to make it happen, it is difficult to descend into deep stages of sleep. In terms of REM sleep, forget about it. Remember that REM sleep is typically accompanied by paralysis. That paralysis can make it significantly harder to keep one's airway open at night because of the reduced airway muscle tone, so REM sleep is usually severely affected by sleep apnea.

THE UNDERWATER REEF EXPLORATION EXERCISE

1. Find a friend. Fly to Cozumel!
2. Both of you need to put on a bathing suit and charter a boat to take you to a deep reef.
3. Outfit yourself with some scuba gear and give your buddy a snorkel and some goggles.
4. Both of you jump in the water.
5. Tell your friend that you are going to explore a deep reef and that it will blow her mind when she sees it.
6. Swim down to the reef.
7. Notice how it is very easy for you to go down to the reef and look at all of the pretty fish and coral. Notice too how your friend starts to descend, but as her oxygen runs out, she must quickly reverse course and swim back to the surface. After grabbing some air, she tries again, but success is fleeting.
8. After your amazing time down at the reef, ascend, collect your friend and head back to shore.

The "Underwater Reef Exploration Exercise" is sleep apnea in a nutshell. Like the snorkeler, the brain desperately wants to descend into the blissful peace of a deep sleep but unfortunately cannot. It simply must wake up to catch its breath. Over and over and over . . .

Why exactly does this happen?

When you stop breathing, several things happen. Levels of oxygen within your body start to fall. When you go to the doctor and she puts the little red light on your finger, she is measuring oxygen levels in your blood. In addition to oxygen levels dropping, carbon dioxide levels rise, as you are not breathing to expel this waste gas.

Your brain is constantly monitoring these levels of oxygen and carbon dioxide, in an effort to maintain balance within your body. When the balance is disturbed from sleep apnea, the body employs a strategy for helping ensure your respiration and continued survival: Your brain essentially scares you awake so you start breathing. Think of all the other things that go along with being scared: rapid heart rate, anxiety, a bump in blood pressure. Yep, they are there too!

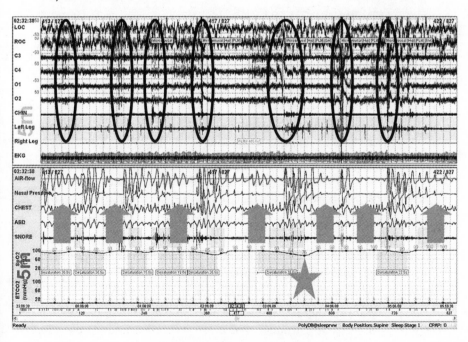

Sleep study showing sleep apnea.

Look at this five-minute section from a sleep study. Do you see the nice up-and-down tracing next to the words "Air-Flow" and "Nasal Pressure"? Great. You're reading your first sleep study. Notice how that up-and-down breathing pattern seems to get a lot less up and down around areas where the arrows are located. In fact looking at some of those arrows, you wonder if the patient is breathing at all. He's not.

These are called apneas (when the patient fully stops breathing and moves no air) or hypopneas (when the patient is moving a little air, but not enough to keep his oxygen level from crashing). Speaking of his oxygen, let's take a look at it. It's labeled "Spo$_2$." Wow, it's a roller coaster! Ideally it would be a boring straight line at or about 98 percent all night long. Look directly above the star. His oxygen gets down to 78 percent! That's not good and in many patients, it gets way worse!

But the fun is only starting! How does snoring play into this suffocation party? Look at the graph marked "Snore." See how the little bursts of activity seem to happen at the end of the nonbreathing periods? That's the individual gasping for breath in a desperate struggle not to suffocate to death. The patient is usually unaware of the issue, although when sleep apnea gets really bad, some patients feel themselves unable to breathe. The bed partner or the person you are sharing the hotel with or the other guys in the hunting cabin are *well* aware of your breathing issues and want to put a pillow over your face.

Finally, take a look at the ovals on top. This section of the sleep study is the EEG . . . the brain wave section. Notice how the EEG or brain activity suddenly changes from the quiet outside of the ovals to bursts of activity within the ovals. During these bursts of activity within the ovals, the brain is waking up. Is this person waking up long enough to remember waking up? Usually not, but not remembering waking up does not mean it's not happening. It is, and it's murder on sleep quality. That's why these people have unbelievable amounts of sleepiness the next day . . . They really didn't sleep!

So let's put this picture all together so we can get a sense of what is happening in sleep apnea:

1. Go to bed.
2. Fall asleep immediately.
3. Begin snoring. Bad for your partner or anyone else in the family, but good for you because at least you are breathing and getting oxygen to your brain.
4. Commence apnea as your airway collapses, preventing breathing.
5. The oxygen in your blood starts to fall.
6. Your brain panics because it is helplessly addicted to oxygen. Your brain says, "Sleep be damned. Wake up. I need to breathe!"
7. A loud snore heralds the return of breathing. You start to breathe again and oxygen levels rise.
8. You fall asleep again and . . . (go to step 3 and repeat).

This process, repeated over a period of months, years, and decades in some cases, takes a heavy toll on your body. I like to think of sleep apnea like rust. If you find a tiny little spot of rust on your car, no big deal. Is it a big deal if you don't get around to sanding and painting until next week? Not at all. Leave it there for the next two years and you might be looking at a very expensive repair.

That's how sleep apnea affects your body. Ignore it for a little while, probably not a big deal. Ignore what your wife, doctor, and friends have been telling you to do for years and you are shortening your life and worsening its quality.

The research is there. In fact, it's everywhere. Sleep apnea affects blood pressure; weight; blood sugar/diabetes; mood/depression; and risk for heart attack, stroke, heart failure, atrial fibrillation, and dying in general. Summary: Sleep apnea is killing you slowly. Believe it.

A final note: Many patients with sleep apnea awaken with headaches and visit the bathroom constantly throughout the night. Take a guess what a patient who has never been diagnosed with sleep apnea gets when he shares the symptoms of headache and frequent urination with his doctor. A sleep study to determine if he has sleep apnea? Ha . . . no way! He gets pills; if he's lucky, several. If he's really lucky, those pills will not be on his insurance formulary and he'll pay through the nose for them.

Don't fall into that trap. If you awaken every morning with a dull headache and you seem to pee much more at night than you do during your waking hours, tell your doctor to evaluate you for sleep apnea.

Sleep Apnea Treatments

Sleep apnea treatments are geared toward eliminating the patient's breathing obstruction. Continuous positive airway pressure (CPAP) is the most common treatment used. CPAP is basically an airway splint that uses air to prop the airway open. It is the most common treatment and probably the most feared.

CPAP devices were developed in the early 1980s by an Australian physician named Colin Sullivan. As if in a dream, he had an idea of how to treat his patient's obstructive breathing events at night. His vision involved taking a Jacuzzi motor apart, and with some tubing and a makeshift mask glued to the patient's face, the pressure generated would hold his airway open and prevent his breathing disturbances at night.

Not only did it work, but it worked so well that even today, the CPAP is still considered the gold standard of care for obstructive sleep apnea syndrome. Advances in technology have made the devices smaller, more comfortable, and more effective. Today, these devices can set their own pressure and even wirelessly provide information about the user's sleep at night.

Besides CPAP, there are other treatments for obstructive sleep apnea. In some cases, sleeping on your side can help. So can weight loss because it often reduces the pressure on your airway. Oral appliances, usually made by dentists, create more airway space by moving the jaw forward in a Marlon Brando, Godfather kind of way. By moving the jaw forward, the tongue is moved out of the airway, creating a larger, more stable opening.

Surgery is another treatment option. It can be as basic as a tonsillectomy or as advanced as breaking the jaw and resetting it or inserting plastic pieces in the palate to keep it from collapsing. A newer surgical option involves implanting a device that stimulates the nerves controlling the muscles that keep the airway open at night. It is still somewhat experimental but may provide a means for patients refractory to other treatments in the future. Other procedures that could be considered involve lasers or an ultrasound designed to shrink the part of the tongue obstructing the airway.

CHAPTER 14 REVIEW

1. Sleep apnea and snoring are not the same thing.
2. Sleep apnea can lead to bad health consequences (heart attacks, stroke, hypertension, heart failure, diabetes, and car wrecks) as well as excessive sleepiness.

Sleep apnea is big-time and it's everywhere. After many, many years, people are finally starting to pay attention to this devastating sleep disorder. It's not the only sleep disorder though. There are many other conditions that can negatively affect sleep. Let's take a closer look at some of the other conditions that can lead to poor sleep at night.

OTHER SLEEP CONDITIONS SO STRANGE, THEY MUST BE SERIOUS

SEE PATIENTS ALL THE TIME in my clinic who say things like this: "I had a sleep study, and it said I didn't have sleep apnea so the doctors don't know what to do to figure out why I keep falling asleep on my tractor."

Sleep apnea is a major contributor to the sleepiness we see during church services across this great land. It is not, however, the only thing that causes excessive daytime sleepiness. Nonetheless, many doctors and some sleep labs treat it as if it were. In other words, many health professionals act like:

No sleep apnea = normal sleep

This could not be further from the truth. You don't have sleep apnea, you say? Congratulations! You don't have to wear a Darth Vader mask. Okay, we've checked off the sleep apnea box. What about the many other sleep diagnoses that exist?

Why would a sleep center pay so much attention to sleep apnea and in some cases ignore other sleep diagnoses? It's usually one of several reasons:

1. Their sleep lab chooses to focus on sleep apnea. This is more common in pulmonary sleep labs (or labs run by pulmonologists—that is, lung doctors), but not exclusive to them. This is one reason to do a little legwork and make sure you are being seen at a true sleep center that deals with all sleep diagnoses, and not a sleep apnea lab.

2. Sleep apnea makes money . . . at least it did when I wrote this book. Things change quickly, but at least as I type, this is a condition that insurance companies will reimburse for. Even now, they are trying to force doctors to use more home studies to diagnose the problem.

Just be aware that there are many sleep disorders out there that can disrupt an individual's sleep and make him feel sleepy during the day.

Restless Legs Syndrome

You are undoubtedly going to think this next sleep problem is made up, but I assure you it is not. The drug companies are not making it up to sell sleeping pills either, as some patients and even some honest-to-God doctors have suggested. The fact that even *Saturday Night Live* caught wind of the condition and did a skit about "restless penis syndrome" is enough to prove the condition sounds absurd. Despite its title, restless legs syndrome (RLS) is a very real condition that affects many disturbed sleepers.

Imagine finishing your dinner and settling in for an evening of SunChips and *Project Runway*. You're there, Heidi's there, the contestants are there, including the one that's really good but kind of bitchy to the others in the workroom . . . what could be better? You're stretched out on the couch, Heidi's explaining the night's challenge, "you're in or you're out," etc. Wait. This isn't comfortable. You pull your legs up and cross them in a yoga-type position: better. Now, then, what did Tim say? Something about keeping a critical eye. Ugh, what's going on with my legs? They feel so uncomfortable . . . like insects crawling around deep inside your legs . . . like an internal itch you can't scratch! Suddenly all you want to do is move and when you do, you feel better. Now with your legs pulled up, kneeling on the couch, you feel marginally improved, but your mood is fading and your interest in a bunch of people making garments out of carpet and tile samples is fading fast. Twenty minutes later, you are on your feet and pacing around feeling like there are caterpillars moving around deep in your legs. As the night goes on, it becomes very apparent that any time you sit down, your legs come alive with new and excitingly annoying sensations.

With your plans for the night shot, you retreat helplessly to bed, resigning yourself to calling it a night. Unfortunately, this problem isn't one you can sleep off. In fact, lying down and resting makes it much worse and only adds to the night's frustration. Your tired body begs for sleep but your legs don't seem to get the message.

What is this unholy condition that seems to disconnect your legs from the rest of your body? This is restless legs syndrome, a condition that is almost impossible for some patients to describe despite the fact that it can affect them on a nightly basis. Often, they will try to bring it up with their primary care doctor.

"Does it hurt?" the physician might inquire.

"Um, well, not really."

"Is it like a cramp?"

"No, it's not a cramp."

"Is it like your feet are on fire, or maybe numb?"

"No. Jeez, it's really hard to describe."

At this point, your primary care physician will probably lose interest in your problem, which does not seem to be particularly urgent or life-threatening. Exercise or mineral water (the condition is sometimes related to mineral deficiency) might be suggested as you are shuffled out the door, frustrated and slightly embarrassed. While these are perfectly valid suggestions and can be effective for some (exercising three days a week has been shown to significantly help some people with RLS), they are not enough for many. In fact, for some sufferers of the condition, exercise paradoxically makes things worse.

Restless legs syndrome is an unusual condition. Not since Vegemite has something been so horribly unpleasant yet so impossible to describe intelligently. When drug companies advertised for patients in some of the earliest drug trials, enrolled patients were often diagnosed for the first time after reading the recruitment posters hung up in the doctors' waiting rooms. Patients with RLS, who up until entering these studies had no idea what they felt was abnormal, could now put a name with their condition and possibly have a cure in their hands. How many migraine patients or seizure patients are unaware that what they feel is not normal?

With possibly 10 percent of the adult population experiencing RLS, based on a 2015 epidemiological study, why are doctors not doing more to diagnose and treat it? I've often heard people say that it's ignored because it's new. Baloney. RLS has been around a long time. Sir Thomas Willis first made a detailed description of the condition back in 1685.

> Wherefore to some, when being a Bed they betake themselves to sleep, presently in the Arms and Leggs Leapings and Contractions of the Tendons, and so great a Restlessness and Tossings of their Members ensue, that the diseased are no more able to sleep, than if they were in a Place of the greatest Torture.
>
> —THOMAS WILLIS, *The London Practice of Physick*
> (London: Basset and Crooke, 1685)

I have no idea how one *betakes* oneself but I'm pretty sure that a "place of great torture" is nowhere I want to be. Sadly, I've had patients describe their experience with RLS in similar terms, often wishing they could cut their legs off to get some rest at night.

So what's been done about getting the word out about RLS over the last 300 years? Pretty much nothing at all. Doctors are good at ignoring things we don't understand and even better at ignoring things we can't treat.

Today we are understanding and treating RLS all over the place. Among the biggest realizations has been that the legs really aren't the problem. It's the brain.[81] Here's how it works.

---

81 It's always the brain!

Our brain is basically a Jell-O mold of chemicals that get secreted here and there, making things happen. Sometimes when a chemical is released, it blocks itself. Other chemicals can block or enhance it as well. It's an amazingly intricate web of helpful forces and oppositional forces. These forces are responsible for everything our bodies do. This push and pull, for instance, make sure neither sleep nor wakefulness ever gains a permanent upper hand in our lives.

To better prepare you for explaining this intelligently to your Facebook friends (or your primary care doctor), substitute the term *neurotransmitter* for *chemical* and voilà, grow your hair out and stop washing your jeans—you are now operating at a grad student level. There are lots of neurotransmitters in the brain. One that is really important is dopamine. Dopamine is important for lots of reasons. First and foremost, as you saw in Chapter 5, it's the big gun when it comes to pleasure. Without it, fraternity parties would be smoke-free affairs, ending around 9:00 P.M. with everyone going home lucid and sleeping by himself.

What else does dopamine do? It wakes us up. It is released in a very circadian fashion with levels being highest during the day and lowest at night. Perfect, because it works out well that we usually like being awake during the day and sleepy at night. Compare the effects of dopamine with those of melatonin (Chapter 2). Notice how they both work to promote sleepiness at night: declining dopamine and rising melatonin. Notice too how you are understanding the chemical nature of sleep. Aren't you glad you decided to *betake* this book? High five.

Any other role for dopamine? Turns out, yes . . . plenty. One big role is the modulation of muscle activity. If you doubt this role, go spend some time with an individual diagnosed with Parkinson's disease or watch on YouTube some recent work by Michael J. Fox, a tireless and brave proponent of Parkinson's awareness and research.[82] Parkinson's disease is a condition that is caused by a sig-

82 For more information, check out the Michael J. Fox Foundation for Parkinson's Research at www.michaeljfox.org. Be sure to betake your credit card and give generously . . . The donation alone will help you sleep at night.

nificant loss of dopamine activity in the brain. If you are spending some time with a Parkinson's patient, invite him to embark on a brisk walk with you. Before you start, make sure he's been without his medications for several days. Ready? Set. Go! Faster . . . come on, push yourself. Finish strong! Excellent effort . . . now catch your breath. How did you do? Chances are when you look back at the starting line, your opponent will just have gotten out of his chair. Notice how slowly he moves and how little his arms swing as he walks. You may also be aware of a tremor. Maybe he decided to hell with this race and simply fell asleep.

Knowing about dopamine, these things make sense. Lack of dopamine makes him tired, unenthusiastic, possibly even depressed (some antidepressants like Wellbutrin act to increase dopamine).

Back to RLS. Fortunately, restless legs syndrome is a snap to diagnose. You can diagnose it yourself—give it a try. Find a friend. Ask her the following questions:

1. "Friend, do you have uncomfortable feelings in your legs sometimes?"
2. "Does moving your legs, walking, and such help make that feeling better?"
3. "Does sitting still make them worse?"
4. "Does this happen more frequently or with more severity at night?"

If you are getting lots of enthusiastic nods, you might be onto something because patients who describe these four symptoms are highly likely to have RLS. This is an important diagnosis to understand because it's often surrounded by much confusion, particularly the fact that *the diagnosis of RLS does not routinely involve a sleep study.* This is a very critical point because (1) many patients are in need of help for their sleep issues but dread or fear having to undergo a sleep study and (2) if a sleep lab is insisting that one undergo a sleep study to diagnose this problem, find another sleep lab.

Once RLS is diagnosed, there are several medications that are FDA approved to treat the condition. Some work by increasing dopamine levels in the brain.[83] The medications are generally well tolerated and can be highly effective. Like the commercial says, "Talk to your physician."

 CHECK YOUR LEGS

1. If you have a Fitbit or some other fitness tracker, mix things up and attach it to your ankle instead of your wrist. If you don't have one, you can borrow one from someone.

2. This exercise works best if you have been wearing a device for a while because you'll be able to compare your findings with the data from your wrist nights.

3. After several ankle nights, take a look at your findings. Does the device seem to indicate a lot more movement when you wear it on your ankle than when you wear it on your wrist? If so, you may be experiencing periodic limb movements during the night. About 70 percent of RLS patients have these tiny little jerks or kicks of their lower extremities during the night. Just like the breathing disturbances in sleep apnea wake people up, these movements also wake people up, leaving them feeling tired and poorly rested during the day.

4. It is interesting that these movements do not affect the upper extremities as much, so fitness trackers can miss them. If there seems to be a big discrepancy, wearing the device may help with figuring out your problem.

One final comment about RLS. It is very hereditary. Be aware that if your mother sleeps poorly and your sister does too, and neither can sit still when your families get together . . . we may be onto something here.

---

83 Thus these drugs are useful in not only restless legs syndrome but also in Parkinson's disease. This fact can cause concern in RLS patients when they discover that the drug they are using is also a Parkinson's drug. Having RLS does not mean you are developing Parkinson's disease.

Narcolepsy

Because of the television show *Seinfeld,* laughter-induced fainting has sometimes been given the name "Seinfeld syndrome." That's a nice way to describe a real problem. People who tend to lose control of their muscles and fall while they're laughing probably have narcolepsy. Narcolepsy, as we have touched on previously, is a condition of excessive daytime sleepiness in which the individual loses the ability to properly stabilize her wakefulness. In other words, normal individuals are usually pretty much wide awake from the time they get up until the time when they go to bed . . . quite a feat when you think back to that adenosine that's always building up in our brains. Narcolepsy patients often lose their foothold in wakefulness and can quickly slip into sleep or experience aspects of sleep while awake and conscious.

Deep within our brains a chemical called orexin is produced. This chemical helps us maintain our ability to stay awake. Orexin is deficient in patients with narcolepsy. Without orexin around, individuals experience an array of unusual sleep symptoms, all based on an inability to stay awake. The five main symptoms of narcolepsy are as follows:

1. Excessive daytime sleepiness and sudden sleep attacks (100 percent of patients with narcolepsy are sleepy—a strong tendency to fall asleep is central to the diagnosis).
2. Hallucinations as you are falling asleep or waking up. The falling-asleep hallucinations are called hypnagogic hallucinations. The waking-up hallucinations are hypnopompic. I always remember the hypnaGOgic ones are when you *go* to sleep. These hallucinations are typically fairly benign, like a cat walking across the bedroom floor but they often cause patients to have trouble distinguishing reality from dreaming.
3. Cataplexy: suddenly becoming weak with laughter or other strong emotions. The muscles stabilizing the knees or upper arms and shoulders are most often af-

fected. This does not necessarily mean you have to fall over. While this condition is often described by patients as well as witnesses as fainting, it's really not. Fainting usually involves a fall and loss of consciousness related to reduced blood flow to the brain. In cataplexy, consciousness is maintained during the period of sudden weakness. In other words, the fainter will often "lose time" during the event, whereas the cataplexy sufferer is usually aware during the entire ordeal, which usually lasts only seconds to a few minutes. This can be very helpful in differentiating cataplexy from fainting (or syncope) or seizures, both of which typically result in consciousness alteration.

4. Sleep paralysis: awakening and becoming conscious but maintaining the paralysis that accompanies REM sleep for a period of time.

5. Disrupted nocturnal sleep. You might think that narcolepsy patients, given their sleepiness, would be expert sleepers. Unfortunately, they are not, as their nights are often punctuated with frequent awakenings.

I cannot stress enough how incredibly impaired and disabled these people are as they try to navigate their lives while being perpetually sleepy for no reason. Strangely, they often have no idea they are disabled. I guess they think everyone walks around fantasizing about sleep because they do. When a patient with narcolepsy wakes up to start the day she stretches and immediately starts to figure out when she can sleep again.

One of my favorite conversations with a narcolepsy patient involved him asking me, "Remember when you were little and you'd go to the hardware store with your dad, and all you wanted to do was take a bunch of paint cans off the shelf, crawl up there and fall asleep?"

He had me up until the paint-can-shelf thing. I looked at him, and politely said, "I have no idea what the hell you are talking about."

To him, he assumed his experience of always being sleepy was a shared experience. Something most everyone goes through. To him, growing up wanting to sleep in an Ace Hardware was as natural as growing up and discovering the joys of the opposite sex.[84]

But it is not natural. It is not natural to fall asleep during your GRE. It is not natural to feel an overwhelming sense of sleepiness looking at the couch behind your drama instructor as she guides you in terms of blocking for the next scene. It is not natural to lie down in the grass during a track practice and be discovered sleeping by your track coach as he's cleaning up after an afternoon of athletics. These are just a few of the stories we have heard.

A father told me when his daughter was diagnosed with narcolepsy, "This is awful." I looked at him and said, "No . . . it's only awful if such patients never get diagnosed." Without a diagnosis, these individuals start to flounder because their need for sleep slowly squeezes out things like school and time for significant others. Narcolepsy patients can often feel inferior or "dumb." In my experience, these patients are anything but dumb. They are often fairly intense and motivated—they have to be to keep up with all of the non-narcolepsy sufferers around them.

Fortunately, there are several available medications that can reduce the sleepiness these patients experience and even reduce the cataplexy attacks from occurring. While most medications we use for this population are stimulants (Ritalin, Adderall) or wakefulness-promoting drugs (Provigil/modafinil and Nuvigil/armodafinil), one drug, Xyrem, is more similar to gamma-hydroxybutyrate (GHB). GHB can be used as a date-rape drug in high doses. But Xyrem is a remarkably effective drug for narcolepsy patients.

Unfortunately, the misunderstanding and fear about this drug within the medical community often prevent it from getting into the hands of the patients who most need it. Truth be told, this is the biggest reason I spend time speaking on behalf of the drug. There are so many undiagnosed narcolepsy sufferers. The average amount of time to get a diagnosis may be as great as fifteen or twenty years. Once these individuals are diagnosed, they deserve to be offered all approved medications that work. It is the patient's job to decide what is right for her, not the doctor's. The age of paternalism in medicine should be over.

*   *   *

---

84 Or the same sex, or both sexes, or neither sex . . . whatever gets you going.

Currently, there are approximately eighty-five recognized sleep disorders. While an explanation of all of them is beyond the scope of this humble book, there are a few more truly weird but very real sleep disturbances to be aware of.

REM Behavior Disorder

Our brain usually does a good job of paralyzing us during sleep. That's a good thing. When I dream I'm trying to fight off a group of howler monkeys, it's great that my brain thinks to turn off the motor first; otherwise Ames is likely to get an elbow to the nose.

In REM behavior disorder, the signal that creates paralysis in the body never gets sent by the brain. The result is an individual who is free to move and act out his dreams at night.

This condition is important because it can be related to Parkinson's disease. In fact it can be a heralding sign in many cases. I'm not saying this to scare you, but just so you are aware that if Grandpoppy suddenly starts reenacting episodes from the war, do not ignore these signs.

Bruxism/Jaw Clenching

Teeth grinding, or bruxism, is a common complaint seen in sleep clinics. It's interesting that teeth grinding is usually not seen during sleep, but rather in the transitional periods between sleep and wakefulness. Remember the sleep apnea patient's sleep study you so expertly read? All of those places where he wakes up to catch his breath are prime periods for grinding enamel.

Most dentists treat bruxism with mouth guards—a physical barrier between one molar and its mate. Very occasionally medications are used. But finding out the underlying reason why a patient is awakening during the night and treating it can often by itself greatly reduce or eliminate bruxism.

Parasomnias: Sleep Talking (Somniloquy), Sleepwalking (Somnambulism), Sleep-Related Eating, Sleepy Sex

Sleep talking, sleepwalking, sleep eating, and sleep sex represent a group of disorders called parasomnias. These disorders are wildly entertaining and fairly common. Having an occasional episode of sleep talking is probably no big deal and doesn't really constitute a disorder. Screaming out obscenities every night and terrifying your partner is probably worth looking into.

Usually these disorders arise from an awakening from deep sleep. A huge contributor to these types of behaviors is sleeping pills, particularly Ambien. The stories of Ambien nocturnal high jinks are well documented. I've had patients interact with their in-laws totally naked, engage in online chats with friends about horrifyingly inappropriate topics, and awaken eating cooking chocolate and raw fingerling potatoes.

Sleep driving has gotten a lot of attention recently, and why shouldn't it? People are driving around asleep and later have no recollection of what they've done. One of my first cases of sleepy driving was a college student who walked out of her dorm in nothing but tiny shorts and a tank top. She proceeded to get into her car and drive. She drove for a while until she got a bit confused. At that point she pulled over and called her parents, five hours away, and said, "Dad, can you come get me?"

"Sweetheart, it will take me hours to get to you. What's going on? It's 3:00 in the morning. . . . Where are you?"

"Oh, just forget it." She hung up. The police fortunately found her shortly afterward and got her back to school safely. She had no recollection of the event. It almost happened again later the same week.

I do not have any great tips for these kinds of behaviors beyond watch out for sleeping pills and alcohol. Beyond that, this is something you might need to work out with a sleep specialist. The underlying causes of these kinds of behaviors can be tough to determine and usually require a sleep study. If you end up needing a sleep study, it's not a big deal. The last chapter will prepare you for what is to come.

....
CHAPTER 15 REVIEW

1. When it comes to sleep, as with your car, there are lots of things that can go wrong. Consider sleep apnea, but do not focus exclusively on it.
2. If a friend, coworker or you have trouble staying awake in a situation, consider the presence of one of these sleep disorders.

Okay, you're considering it, but you want to know if you have it or not. What to do? Have a sleep study! They are fun and give you the opportunity to make a grainy videotape of you doing something freaky in a hotel bed—worked out well for Kim Kardashian and Paris Hilton! How do you have a sleep study and what can you expect? The finish line is in sight!

16

TIME FOR A
SLEEP STUDY

S LEEP STUDIES ARE AWESOME. MORE uncomfortable than a
tonsil swab for a strep test, but less uncomfortable than a colo-
noscopy. Sleep studies are designed to allow a sleep specialist
to monitor numerous aspects of your sleep, including breathing,
brain activity, and muscle activity. By looking at your sleep, you hope
somebody can figure out what your problem is.

Sleep studies are a great tool in your doctor's handyman box for
helping fix you up, but they have limitations and in some cases are
not necessary or helpful.

Keep in mind, a thirty-year-old who averages about seven hours
of sleep at night has slept approximately 76,650 hours in her lifetime. Thus a one-night sleep study represents a 0.00009 percent
sample of this person's lifetime sleep. It's a very small sample. However, in the right situation, this tiny sample can hold the key to an
individual's sleep problems.

Many people fear sleep studies. The whole arrangement just
seems a little strange. It doesn't help that most sleep centers have
rooms that are sterile . . . like the room in the spaceship you would
envision being brought to after your alien abduction. But more
and more sleep centers have their eye on patient comfort, and the

whole setup can be almost posh. In some instances, sleep studies can even be done in the patient's own home.

The Inpatient Sleep Study

What's involved in a sleep study? In a word, glue. Brace yourself, because after your sleep study you are going to be picking and scrubbing little bits of glue out of your hair and from behind your ears for days. If I don't tell a patient about the glue beforehand, I'm certain to hear about it when he returns for the results. I've learned to warn people in advance.

That glue is important because it firmly attaches little wires to patients during the night. These wires (or *leads*) measure tiny electrical impulses given off by the brain or muscles. This, combined with devices that measure breathing, the oxygen content of your blood, and heart rate, constitutes a sleep study, or polysomnogram.

Sleep studies combine these elements into a continuous picture of a person's sleep. At the study's foundation is the level of sleep, or the staging. Remember how sleep can be divided into three stages: dream sleep, light sleep, and deep sleep? Well, this is where

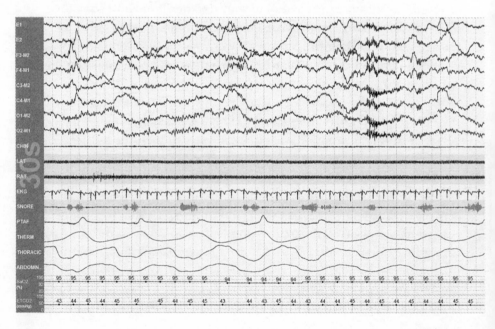

we actually see those stages happen—live—during the night. So cool.

I know what you're asking. How can you tell the different stages of sleep apart? It is surprisingly easy. By paying attention to the brain wave activity (EEG), the eye movement (EOG), and the muscle activity (EMG), we can determine the stages in a flash.

The figure on the previous page shows a lot of little wiggly lines. What exactly are you looking at there? A lie detector test? Close, but no. This is a screenshot of a sleep study. All of those lines are outputs from the leads attached to the patient. Here is what they are measuring, item by item.

Eye Movements

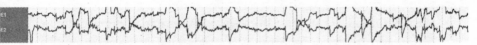

This tracing taken from a sleep study shows the movement of the LOC (left eye) and the ROC (right eye). Because of the way the electrodes are placed on the face, it looks like the eyes are moving in different directions. They are not. Eyes move around a lot while an individual is awake, and less so as he sleeps. During REM sleep (or rapid eye movement sleep), guess what they do. That's right: they move rapidly. During deep sleep they move very little. Already, we see a way that we can start to distinguish different stages of sleep.

Brain Wave Activity (EEG)

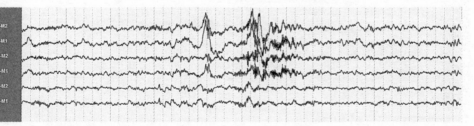

A central component of the sleep study is the measurement of your brain wave activity. Pay attention to the examples of the different stages of sleep and notice the differences yourself. Some waves look big; others look small. Some waves look rapid and others, slow. When an individual is awake, as in the example given here, notice how quickly the waves go up and down and how short they are. Sleep studies usually begin and end with the subject awake. In addition, patients awaken during their studies (sometimes frequently—trying to figure out why is often the reason for having the study). Recognizing wakefulness is an essential component of reading a sleep study.

Muscle Activity

We measure muscle activity from three places typically: the chin, left leg (Note: *LAT* stands for *left anterior tibialis*. This muscle is responsible for lifting your foot, and it's the muscle typically measured), and right leg (RAT). Muscle tone is high when you are awake, less in light sleep and dream sleep, and gone when you dream.

Combining these three measurements, it's pretty clear how we determine what stage of sleep you are in.

	WAKE	LIGHT SLEEP (Stages N1/N2)	DEEP SLEEP (Stage N3)	DREAM SLEEP (REM)
Eye Movement (EOG)	Lots, plus loads of blinking!	Less . . . slow and rolling	None	Rapid eye movements seen (hence the name REM)
Brain Waves (EEG)	Fast and short	Slower and a bit taller	Very slow and really tall	Fast and short, similar to Wake
Muscle Activity (EMG)	Lots	Less	Less	None

What could be easier? Just look at what the polysomnogram is showing and you can figure out what stage of sleep you are looking at. Let's put it to the test!

VIRTUAL SLEEP DOCTOR EXERCISE

"DOCTOR [YOUR NAME], WE NEED your help interpreting this sleep study finding. We are involved in an incredibly complex case and lives hang in the balance. Can you tell us what stage of sleep this is?"

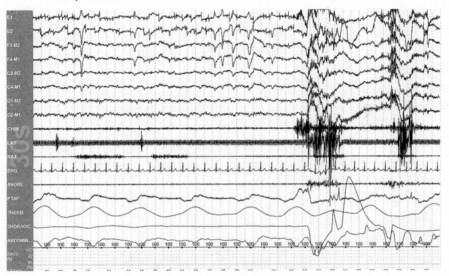

1. Notice all of those eye movements?
2. Notice that fast and short EEG?
3. Notice all of that muscle tone?
4. Make your guess!

You guessed it: This patient is awake. I know so because as I recorded this I asked the patient, "What's up?" He said, "Nothing much." I really didn't care what was up with this man. What I was interested in was whether or not he was awake. Look at that fast EEG. His eyes are checking out the sleep center as they roam all around. His muscle tone is high. Notice too beside the labels "Air Flow," "Chest" and "Abdomen" how nice his breathing . . . inhale, exhale, inhale, exhale . . . looks.

Light Sleep

As we learned in Chapter 4, light sleep forms the foundation of a night of sleep (about 50 percent). So how do we recognize light sleep? Take a look at the example sleep study. Notice the relatively low height of the first eight lines. These are the eye movement and brain wave activity recordings. Light sleep features a relatively low amplitude (height) of brain waves and eye movement activity. If you also recall from Chapter 4, there are two categories of light sleep: N1 and N2. Differentiating the two is done by looking for two distinct features called *sleep spindles* and *K-complexes*. Sleep spindles and K-complexes are characteristic of N2 sleep, so when you see them, you have moved from N1 into N2 sleep. Before we leave light sleep, take a look at the channel labeled "Snore." Do you see the intermittent periods of snoring? Notice how they relate to the up-and-down movement of the breathing channels immediately above and below.

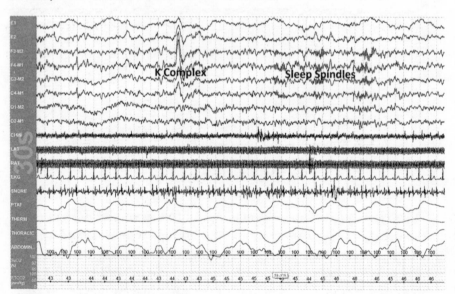

Deep Sleep

Now we get to the good stuff. Deep sleep, or N3 sleep, is the sleep that really refreshes us the next day. Contrast this example with the one you just studied. First, look at the top six lines—the brain activity. Notice how much taller and broader these waves are than the ones seen in light sleep. Look at the breathing ("Air Flow," "Chest," "Abdomen"). Do you notice how perfectly regular the breathing seems to be? In N3 sleep, the thinking part of the brain is in its most relaxed state with the more primitive parts of the brain running the show.

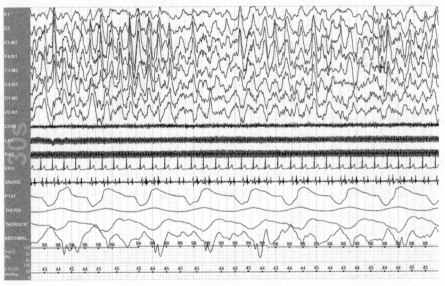

REM Sleep

Finally we take a look at dreaming. In this example, you can see how this stage of sleep got the name "rapid eye movement." Look at the massive waves in the first two eye channels (E1 and E2). These waves are produced by our eyes moving while we dream. Our eyes escape the muscle paralysis usually seen in REM sleep. The paralysis that affects the rest of the body is clearly seen by looking at the flat lines of the three muscle activity channels ("Chin," "RAT," "LAT").

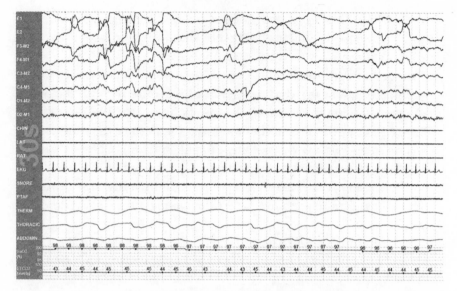

These examples are meant just to serve as a glimpse as to what sleep specialists are looking at when a sleep study is done. No scary shots or needles. No creepy anal probes or other such sinister devices. Just some wires and glue, and a video camera, so wear something nice.

In most cases, you are going to arrive for your sleep study around 8:00 P.M. Sometimes a sleep center is located in a hospital (good luck sleeping through the cardiac arrest code). Other times the study is in a remote location. Our sleep center, like many others, is actually in a nice hotel. The more comfortably you sleep, the better the results.

Regardless of the location, you are going to show up with your jammies and toothbrush ready to go. You'll be met by a technologist who will escort you to your private room and make you feel comfortable. At some point, when you are dressed and ready to go to sleep, the technician will return to hook you up for the study. This basically involves taping or gluing little wires to different parts of your body (told you there was glue). Once you are hooked up, you can move about in the bed as you like; you can even get up to go to the bathroom quite easily. Most of the wires attached to you feed into one small box that is then plugged in next to your bed. If you want to get up, a technician simply unplugs that box from next to your bed. Easy! Don't worry. You will not be tied down to the bed. A final note. Anytime you need something, all you have to do is speak out loud and the technician will hear you. Keep in mind that the technician is monitoring many things about you during the night including your brain activity. Chances are, the technician will know what you need before you do!

When you are ready to sleep, you can turn the lights out and sleep. Don't worry. You don't have to have a perfect night of sleep for the sleep study to diagnose the problem. You don't even need to have a good night of sleep . . . a few hours will do. If you are concerned about achieving sleep during your study, stay up late the night before so you are a little extra sleepy. That should do the trick.

You can move around in bed during a sleep study and sleep in whatever position feels comfortable to you. You are always free to bring your own pillows, blankets or other things from home that make you feel comfortable. As you sleep, don't be afraid a wire may come off . . . the technicians are prepared to handle that situation if necessary.

When the morning comes, you are typically free to leave whenever you awaken. The technician will help you remove the wires from your body and clean off the paste used to hold them in place. For many patients, removing the little bits of glue from their scalp is the only difficult part of the whole affair!

Home Sleep Testing

Within the last few years, there has been a massive shift toward home sleep testing, or HST. A home sleep test is a portable device that allows sleep doctors to look at your sleep from the comfort and privacy of your own bed. While being able to sleep in one's own bed is a real positive, home sleep studies have their drawbacks and limitations. Understanding the differences between in-lab sleep studies and home sleep testing, and choosing the right test, is essential for getting to the bottom of your sleep problem.

So what is a home sleep test? It is basically a simple device that the patient puts on and wears during the night in their own bedroom. No doctor, no sleep technician, just you and some wires!

The most commonly used devices today typically monitor five biological outputs:

1. Airflow or air pressure (at the nose, mouth, or both)
2. Breathing effort (at the chest most commonly)
3. Oxygen saturation
4. Pulse
5. Snoring

While this is a dazzling array of things to monitor, do you notice anything missing? Remember all of the things we needed to record in order to determine whether or not someone was sleeping and what stage of sleep they are in (eye movements, brain activity, muscle tone)? Well, it turns out that most home studies record none of those things! In other words, the one thing a home sleep study does not study is sleep! Because of this, many sleep doctors abhor the term *home sleep study* because it is largely untrue.

Think about that. The HST does not record sleep. It's really a home-breathing study and is equipped only to answer the questions: Does my patient breathe? Does my patient snore? Does my patient have a heartbeat? I can figure out the answer to two of those questions before the patient sits down to read *Rolling Stone* in my waiting room.

Who cares? Details and semantics, right? Not really, because to determine whether or not an individual has sleep apnea, we need to know how much that individual slept at night so we can work our sleep apnea equation:

Number of Breathing Problems ÷ Time Asleep = Number of Breathing Problems per Hour of Sleep

Now you can see why the home sleep study has a big problem. The home sleep study typically does not measure sleep, so when the data of a home sleep study are analyzed, the doctor cannot determine if the patient ever actually sleeps.

In the place of sleep time is test time, or how long the device is on as it is being worn by the patient. This is great if the patient falls asleep immediately and does not awaken until the study ends, but this is seldom the case—especially if the patient actually does have sleep apnea. The end result is, the more a patient is awake during the study (and not having breathing disturbances), the more these devices underestimate the severity of a patient's sleep apnea. Because breathing disturbances do not happen while a patient is awake, the time spent awake and not struggling to breathe is logged as normal-breathing sleep time.

That is not the only issue with these devices. They are also prone to be manipulated. Imagine you are a truck driver who knows that if you have a sleep disturbance, your commercial driver's license might be in jeopardy. When your doctor hands you a home sleep testing device, letting your wife wear it to ensure a normal report might seem like a great idea. Believe me, it happens.

Home sleep studies have their place. For someone who clearly has sleep apnea, this choice can lead to significant cost savings. They were really designed for use in the following patients:

1. Patients likely to have sleep apnea or a breathing disturbance.
2. Patients unable to have a regular sleep study because of lack of insurance or inability to easily leave the home for health or social reasons (a sole caregiver), or people who for other reasons can't spend a night away from home.

The biggest issue with HSTs is the way in which insurance companies have dictated their use. If there is a good chance that you have sleep apnea, this is probably a good test. If you are a twenty-two-year-old woman who does not snore, has felt sleepy for years, and acts out your dreams, tell your insurance company you need an inpatient sleep study. The home study is going to be as helpful as a prostate exam in her quest to repair her sleep.

A final word about sleep studies. Insist that when the study is complete, you are able to sit down with the doctor who interpreted the study and go over it with him or her. Your insurance company may have just spent close to two grand for you to have this experience. You deserve to have the test results explained to you by someone who understands sleep. It's great if you have a primary care doctor who thinks about sleep and is thoughtful enough to order a sleep study. You still deserve to have the test explained to you beyond the doctor simply reading the interpretation.

....

CHAPTER 16 REVIEW

1. Sleep studies are really helpful and not something to fear.
2. Sleep studies collect a tremendous amount of data on sleep.
3. If you are forced to do a home sleep study to diagnose your sleep disturbance and it fails to do so, insist on your insurance company paying for a real inpatient sleep study. I would also suggest that this study be ordered by a sleep specialist, not by your primary care physician.

See, with a sleep study there is nothing to fear. What's more, after reading this chapter, you are going to be primed to sit down with your sleep specialist and go over your sleep study in an educated and prepared way.

Speaking of sitting down with your specialist, many patients I encounter who have had sleep studies in the past say they never had the chance to speak to a sleep specialist about their study or even to get the results of the study. Many say that they were told that the study was "inconclusive" or "normal" and received no further help.

Every sleep study provides useful information. Insist on going over the results of your study. It is an essential part of understanding and treating your sleep problems. Remember, this study should mark the beginning of your sleep disturbance treatment process and the relationship with your sleep professional, not the end.

Conclusion

S THIS BOOK A COMPLETE reference for the subject of sleep? No, it's not, and it is not supposed to be. There are other books for that. Bill Dement's *The Promise of Sleep* is a wonderfully detailed exploration of lots of sleep disorders if you want a better explanation than what you can find on Wikipedia.

This book provided you something more than that. I hope it has helped mold a broader view of sleep that will allow you to better identify what is going on with your sleep (if anything) and how to fix it.

I started writing this book while sitting in Hartsfield Airport waiting for a connecting flight home to Charlottesville. It began as an exercise to record the things I talk to patients about in my clinic—a way of archiving the explanations and techniques that work. The stories and analogies in this book have been honed over years of observing other doctors and practicing with my own clinic patients.

Dealing with sleep problems is hard, as it is often very difficult to get a perspective on someone's sleep. It's like trying to pluck a hair out of the middle of your back. First, it's difficult to see if the hair is even there. To do so would probably involve holding a mirror while looking at another one—very tough. Second, even once you've identified the renegade hair, plucking it out yourself is probably bordering on the impossible. This book has provided you some enhanced insight into your own sleep (as if I held the mirror for you), while at the same time it has given you some better ideas of how to pluck it.

I have one departing wish for you, Reader, and it came about via the comments given to me by the first real editor to read my book. Deep into the book, while I was explaining something about insomnia, the editor wrote, "Is this sentence helpful to people who suffer from insomnia so bad that it's not just the occasional bump in the road?" I was crestfallen. Had everything I had written failed to show my editor that this person suffering so badly was no different from the person having occasional insomnia? Had the whole idea of the insomnia mind-set, so to speak, evaded her? After thinking about it for a while, and reflecting on the thousands of patients I've seen over the years, I came up with a final thought:

Achieving great sleep can take time.

Building a muscular body when you are a bit overweight and out of shape takes time. Learning conversational Italian takes time. Nothing that is great comes quickly, and I'm afraid sleep is no different. So if you read my book and are sleeping the best sleep of your life by the time you reach this section, I could not be more thrilled. If not, my suggestion is to take some time to digest what you've learned. Try out some of the things I've suggested. You might find with time, you'll find in this book the solutions you need. That's my hope.

Author's Note

HAVE NO FINANCIAL RELATIONSHIP WITH any of the products in this book. These are products I have discovered over many years of helping people sleep better, so rest assured your purchases of these products are not funding my vintage sleeping cap collection. I have served as a paid consultant/speaker for various restless legs syndrome and narcolepsy medications because I feel strongly that many doctors need help recognizing and appropriately treating these conditions. Despite numerous invitations, I have never given a paid lecture for a sleeping pill.

Acknowledgments

MY ROAD TO BECOMING A sleep specialist was paved by three amazing physicians I had the fortune of working with and learning from over the years. I want to take this opportunity to thank those gentlemen.

Paul Suratt, the former head of sleep at the University of Virginia, was more than a mentor to me. He is a friend and brilliant role model. He gave me my first exposure to sleep as an undergraduate student and showed me how amazing this field can be. You are reading this book because of him. Thank you, Paul.

Paul introduced me to Don Bliwise when I went to Emory University medical school. He runs the Emory University Sleep Center. If Paul provided the spark, Don made the fire. I cannot imagine a kinder man who has bent over backward to help me ever since. He is generous with his time and universally adored. Thank you, Don.

Finally, to make things official, I did a sleep medicine fellowship with Brad Vaughn, the head of sleep at UNC–Chapel Hill. Brad taught me the nuts and bolts of how to be a sleep center director, and what he didn't teach me, I simply plagiarized from him. Think about the hardest worker you know and Brad did that person's day's work during his lunch break.[85] While I'll never live up to your work ethic, I at least have a target to shoot for. Thank you, Brad.

I'd like to thank Justo Campa for inviting me to join his practice many years ago and entrusting me with it after his retirement. I want to thank everyone in my office: Perri, Geni, Betsy, Sharon, and Johanna for making my work life so fun. I want to thank

---

85 Kidding . . . Brad would never take a lunch break.

Tammy especially for managing my hectic life so gracefully and being my rock. You are an exceptional office manager and anyone who tries to steal you from my office is going to find themselves in a scene right out of *Goodfellas*.

Thank you to Jeff for deftly guiding this book to completion. The book had become this strange imaginary friend. It lived on my computer, but nobody could see it despite the fact that I referenced it from time to time. You truly are responsible for it coming to life.

Thank you to Claire Zion and the wonderful folks at Penguin. You really took a chance here, and your support was evident from the moment we met in New York. Free sleep advice forever.

Thank you, David Bowie. I had a dream that one day you'd call me out of the blue[86] about this troubling dream you were having about floating around in space, and I would help you sort things out. I'm sad that won't happen. I love your music.

Most important, thank you to my family for being so supportive of this project. Maeve, Tyce, and Cam, you are not only the best sleepers in the world, but you are pretty nice young people to be around. To my wife, Ames, who never stopped asking me, "When are you going to do something with that book?" I dedicate this to you.

---

86 Electric blue.

Bibliography

AN INTRODUCTION TO SLEEP MEDICINE

1. Roth, T. "Insomnia: Definition, Prevalence, Etiology, and Consequences." *Journal of Clinical Sleep Medicine* 3, suppl. 5 (2007): S7–S10.
2. Ohayon, M. M., R. O'Hara, and M. V. Vitiello. "Epidemiology of Restless Legs Syndrome: A Synthesis of the Literature." *Sleep Medicine Reviews* 16, no. 4 (2012): 283–95.
3. National Sleep Foundation. *2005 Sleep in America Poll Summary of Findings*. Washington, DC: National Sleep Foundation, 2005.
4. Rosen, R. C., M. Rosenkind, C. Rosevar, et al. "Physician Education in Sleep and Sleep Disorders: A National Survey of U.S. Medical Schools." *Sleep* 16, no. 3 (1993): 249–54.
5. Teodorescu, M. C., A. Y. Avidan, M. Teodorescu, et al. "Sleep Medicine Content of Major Medical Textbooks Continues to Be Underrepresented." *Sleep Medicine* 8, no. 3 (2007): 271–76.

CHAPTER 1

1. Louveau, A., I. Smirnov, T. J. Keyes, et al. "Structural and Functional Features of Central Nervous System Lymphatic Vessels." *Nature* 523 (2015): 337–41.
2. Aspelund, A., S. Antila, S. T. Proulx, et al. "A Dural Lymphatic Vascular System That Drains Brain Interstitial Fluid and Macromolecules." *Journal of Experimental Medicine* 212, no. 7 (2015): 991–99.
3. Xie, L., H. Kang, Q. Xu, et al. "Sleep Drives Metabolite Clearance from the Adult Brain." *Science* 342, no. 6156 (2013): 373–7 .
4. Spira, A. P., A. A. Gamaldo, Y. An, et al. "Self-Reported Sleep and ß-Amyloid Deposition in Community-Dwelling Older Adults." *JAMA Neurology* 70, no. 12 (2013): 1537–43.
5. Lim, A. P., L. Yu, M. Kowgier, et al. "Sleep Modifies the Relation of *APOE* to the Risk of Alzheimer Disease and Neurofibrillary Tangle Pathology." *JAMA Neurology* 70, no. 12 (2013): 1544–51.
6. Lee, H., L. Xie, M. Yu, et al. "The Effect of Body Posture on Brain Glymphatic Transport." *Journal of Neuroscience* 35, no. 31 (2015): 11034–44.
7. Suzuki, K., M. Miyamoto, T. Miyamoto, et al. "Sleep Disturbances Associated with Parkinson's Disease." *Parkinson's Disease* 2011 (2011): 10 pages.
8. Schönauer, M., A. Pawlizki, C. Köck, and S. Gais. "Exploring the Effect of Sleep and Reduced Interference on Different Forms of Declarative Memory." *Sleep* 37, no. 12 (2014): 1995–2007.

9. Baron, K. G., K. J. Reid, A. S. Kern, and P. C. Zee. "Role of Sleep Timing in Ca-
    loric Intake and BMI." *Obesity* 19, no. 7 (2011): 1374–81.
10. Patel, S. R., and F. B. Hu "Short Sleep Duration and Weight Gain: A Systematic
    Review." *Obesity* 16, no. 3 (2008): 643–53.
11. Zhang, J., X. Jin, C. Yan, et al. "Short Sleep Duration as a Risk Factor for Child-
    hood Overweight/Obesity: A Large Multicentric Epidemiologic Study in China."
    Sleep Health 1, no. 3 (2015): 184–90.
12. Sperry, S. D., I. D. Scully, R. H. Gramzow, and R. S. Jorgensen. "Sleep Duration
    and Waist Circumference in Adults: A Meta-Analysis." *Sleep* 38, no. 8 (2015):
    1269–76.
13. Van Cauter, E., and K. L. Knutson. "Sleep and the Epidemic of Obesity in Chil-
    dren and Adults." *European Journal of Endocrinology* 159, no. S1 (2008): S59–S66.
14. Taheri, S., L. Lin, D. Austin, et al. "Short Sleep Duration Is Associated with Re-
    duced Leptin, Elevated Ghrelin, and Increased Body Mass Index." *PLoS Medicine*
    1, no. 3 (2004): e62.
15. Hakim, F., Y. Wang, A. Carreras, et al. "Chronic Sleep Fragmentation During the
    Sleep Period Induces Hypothalamic Endoplasmic Reticulum Stress and
    PTP1b-Mediated Leptin Resistance in Male Mice." *Sleep* 38, no. 1 (2015): 31–40.
16. Lundahl, A., and T. D. Nelson. "Sleep and Food Intake: A Multisystem Review of
    Mechanisms in Children and Adults." *Journal of Health Psychology* 20, no. 6 (2015):
    794–805.
17. Killgore, W. D. S., T. J. Balkin, and N. J. Wesensten. "Impaired Decision Making Fol-
    lowing 49 Hours of Sleep Deprivation." *Journal of Sleep Research* 15, no.1 (2006): 7–13.
18. Asarnow, L. D., E. McGlinchey, and A. G. Harvey. "Evidence for a Possible Link
    Between Bedtime and Change in Body Mass Index." *Sleep* 38, no. 10 (2015):
    1523–27.
19. Kanagala, R., N. S. Murali, P. A. Friedman, et al. "Obstructive Sleep Apnea and
    the Recurrence of Atrial Fibrillation." *Circulation* 107, no. 20 (2003): 2589–94.
20. Luca, A., M. Luca, and C. Calandra. "Sleep Disorders and Depression: Brief Re-
    view of the Literature, Case Report, and Nonpharmacologic Interventions for De-
    pression." *Clinical Interventions in Aging* 8 (2013): 1033–39.
21. Finan, P. H., P. J. Quartana, and M. T. Smith. "The Effects of Sleep Continuity
    Disruption on Positive Mood and Sleep Architecture in Healthy Adults." *Sleep* 38,
    no. 11 (2015): 1735–42.
22. Edwards, C., S. Mukherjee, L. Simpson. "Depressive Symptoms Before and After
    Treatment of Obstructive Sleep Apnea in Men and Women." *Journal of Clinical
    Sleep Medicine* 11, no. 9 (2015): 1029–38.
23. Jindal, R. D., and M. E. Thase. "Treatment of Insomnia Associated with Clinical
    Depression." *Sleep Medicine Reviews* 8 (2004): 19–30.
24. Markt, S. C., A. Grotta, O. Nyren, et al. "Insufficient Sleep and Risk of Prostate
    Cancer in a Large Swedish Cohort." *Sleep* 38, no. 9 (2015): 1405–10.
25. Fang, H. F., N. F. Miao, C. D. Chen, et al. "Risk of Cancer in Patients with Insom-
    nia, Parasomnia, and Obstructive Sleep Apnea: A Nationwide Nested Case-Con-
    trol Study." *Journal of Cancer* 6, no. 11 (2015): 1140–47.
26. Zhang, X., E. L. Giovannucci, K. Wu, et al. "Associations of Self-Reported Sleep
    Duration and Snoring with Colorectal Cancer Risk in Men and Women." *Sleep* 36,
    no. 5 (2013): 681–88.
27. Chen, J. C., and J. H. Hwang. "Sleep Apnea Increased Incidence of Primary Cen-
    tral Nervous System Cancers: A Nationwide Cohort Study." *Sleep Medicine* 15, no. 7
    (2014): 749–54.
28. Wang, P., F. M. Ren, Y. Lin, et al. "Night-Shift Work, Sleep Duration, Daytime
    Napping, and Breast Cancer Risk." *Sleep Medicine* 16, no. 4 (2015): 462–68.

29. Phipps, A. I., P. Bhatti, M. L. Neuhouser, et al. "Prediagnostic Sleep Duration and Sleep Quality in Relation to Subsequent Cancer Survival." *Journal of Clinical Sleep Medicine* 12, no. 4 (2016): 495–503.

30. Straif, K., R. Baan, Y. Grosse, et al. "Carcinogenicity of Shift-Work, Painting, and Fire-Fighting." *Lancet* 8, no. 12 (2007): 1065–66.

31. Erren, T. C., P. Falaturi, P. Morfeld, et al. "Shift Work and Cancer: The Evidence and the Challenge." *Deutsches Ärzteblatt International* 107, no. 38 (2010): 657–62.

32. Prather, A. A., D. Janicki-Deverts, M. H. Hall, and S. Cohen. "Behaviorally Assessed Sleep and Susceptibility to the Common Cold." *Sleep* 38, no. 9 (2015): 1353–59.

33. Hsiao, Y. H., Y. T. Chen, C. M. Tseng, et al. "Sleep Disorders and Increased Risk of Autoimmune Diseases in Individuals without Sleep Apnea." *Sleep* 38, no. 4 (2015): 581–86.

CHAPTER 2

1. Hull, C. *Principles of Behavior.* New York: Appleton-Century-Crofts, 1943.

2. Van Dongen, H. P., G. Maislin, J. M. Mullington, and D. F. Dinges. "The Cumulative Cost of Additional Wakefulness: Dose-Response Effects on Neurobehavioral Functions and Sleep Physiology from Chronic Sleep Restriction and Total Sleep Deprivation." *Sleep* 26, no. 2 (2003): 117–26.

3. Cirelli, C., and G. Tononi. "Is Sleep Essential?" *PLoS Biology* 6, no. 8 (2008): e216.

4. Cano, G., T. Mochizuki, and C. B. Saper. "Neural Circuitry of Stress-Induced Insomnia in Rats." *Journal of Neuroscience* 28, no. 40 (2008): 10167–84.

5. Ohayon, M. M., M. A. Carskadon, C. Guilleminault, and M. V. Vitiello. "Meta-Analysis of Quantitative Sleep Parameters from Childhood to Old Age in Healthy Individuals: Developing Normative Sleep Values Across the Human Lifespan." *Sleep* 27, no. 7 (2004): 1255–73.

6. Hirshkowitz, M., K. Whiton, S. M., Albert, et al. "National Sleep Foundation's Sleep Time Duration Recommendations: Methodology and Results Summary." *Sleep Health* 1, no. 1 (2015): 40–43.

7. Knutson, K. L., E. Van Cauter, P. J. Rathouz, et al. "Trends in the Prevalence of Short Sleepers in the USA: 1975–2006." *Sleep* 33, no. 1 (2010): 37–45.

8. Yetish, G., H. Kaplan, M. Gurven, et al. "Natural Sleep and Its Seasonal Variations in Three Pre-Industrial Societies." *Current Biology* 25, no. 21 (2015): 2862–68.

CHAPTER 3

1. National Transportation Safety Board. "Grounding of the U.S. Tankship Exxon Valdez on Bligh Reef, Prince William Sound Near Valdez, Alaska. March 24, 1989" [marine accident report] (PB90-916405 NTSB/MAR-90/04).

2. Watson, N. F., M. S. Badr, G. Belenky, et al. "Joint Consensus Statement of the American Academy of Sleep Medicine and Sleep Research Society on the Recommended Amount of Sleep for a Healthy Adult: Methodology and Discussion." *Journal of Clinical Sleep Medicine* 11, no. 8 (2015): 931–52.

3. Johns, M. W. "A New Method for Measuring Daytime Sleepiness: The Epworth Sleepiness Scale." *Sleep* 14, no. 6 (1991): 540–45.

4. Goldstein-Piekarski, A. N., S. M. Greer, J. M. Saletin, and M. P. Walker. "Sleep Deprivation Impairs the Human Central and Peripheral Nervous System Discrimination of Social Threat." *Journal of Neuroscience* 35, no. 28 (2015): 10135–45.

5. Simon, E. B., N. Oren, H. Sharon, et al. "Losing Neutrality: The Neural Basis of Impaired Emotional Control Without Sleep." *Journal of Neuroscience* 35, no. 38 (2015): 13194–13205.

6. Burke, T. M., R. R. Markwald, A. W. McHill, et al. "Effects of Caffeine on the Human Circadian Clock In Vivo and In Vitro." *Science Translational Medicine* 7, no. 305 (2015): 305ra146.
7. Gooley, J. J., J. Lu, D. Fischer, and C. B. Saper. "A Broad Role for Melanopsin in Nonvisual Photoreception." *Journal of Neuroscience* 23, no. 18 (2003): 7093–7106.
8. Flourakis, M., E. Kula-Eversole, A. L. Hutchison, et al. "A Conserved Bicycle Model for Circadian Clock Control of Membrane Excitability." *Cell* 162, no. 4 (2015): 836–48.

CHAPTER 4

1. Alapin, I., C. S. Fichten, E. Libman, et al. "How Is Good and Poor Sleep in Older Adults and College Students Related to Daytime Sleepiness, Fatigue, and Ability to Concentrate?" *Journal of Psychosomatic Research* 49, no. 5 (2000): 381–90.
2. Aserinsky, E., and N. Kleitman. "Regularly Occurring Periods of Eye Motility, and Concomitant Phenomena, During Sleep." *Science* 118, no. 3062 (1953): 273–74.
3. Tilley, A. J., and J. A. Empson. "REM Sleep and Memory Consolidation." *Biological Psychiatry* 6, no. 4 (1978): 293–300.
4. Greenhill, L., J. Puig-Antich, R. Goetz, et al. "Sleep Architecture and REM Sleep Measures in Prepubertal Children with Attention Deficit Disorder with Hyperactivity." *Sleep* 6, no. 2 (1983): 91–101.
5. Palagini, L., C. Baglioni, A. Ciapparelli, et al. "REM Sleep Dysregulation in Depression: State of the Art." *Sleep Medicine Reviews* 17, no. 5 (2013): 377–90.
6. Modell, S., and C. J. Lauer. "Rapid Eye Movement (REM) Sleep: An Endophenotype for Depression." *Current Psychiatry Reports* 9, no. 6 (2007): 480–85.
7. Roehrs, T., M. Hyde, B. Blaisdell, et al. "Sleep Loss and REM Sleep Loss Are Hyperalgesic." *Sleep* 29, no. 2 (2006): 145–51.
8. Vanini, G. "Sleep Deprivation and Recovery Sleep Prior to a Noxious Inflammatory Insult Influence Characteristics and Duration of Pain." *Sleep* 39, no. 1 (2016):133–42.
9. Van Cauter, E., and G. Copinschi. "Interrelationships between Growth Hormone and Sleep." *Growth Hormone & IGF Research* 10, suppl. B (2000): S57–62.

CHAPTER 5

1. Gray, S. L., M. L. Anderson, S. Dublin, et al. "Cumulative Use of Strong Anticholinergic Medications and Incident Dementia." *JAMA Internal Medicine* 175, no. 3 (2015): 401–7.

CHAPTER 6

1. An., H., and S. A. Chung. "A Case of Obstructive Sleep Apnea Syndrome Presenting As Paradoxical Insomnia." *Psychiatry Investigations* 7, no. 1 (2010): 75–78.
2. Case, K., T. D. Hurwitz, S. W. Kim, et al. "A Case of Extreme Paradoxical Insomnia Responding Selectively to Electroconvulsive Therapy." *Journal of Clinical Sleep Medicine* 4, no. 1 (2008): 62–63.
3. Ghadami, M. R., B. Khaledi-Paveh, M. Nasouri, and H. Khazaie. "PTSD-Related Paradoxical Insomnia: An Actigraphic Study Among Veterans with Chronic PTSD." *Journal of Injury and Violence Research* 7, no. 2 (2015): 54–58.

CHAPTER 7

1. Kleitman, N. "Periodicity." *Sleep and Wakefulness*. University of Chicago Press, 1963.

2. *The International Classification of Sleep Disorders: Diagnostic and Coding Manual.* Revised. Westchester: American Academy of Sleep Medicine, 2001.
3. Liira, J., J. Verbeek, and J. Ruotsalainen. "Pharmacological Interventions for Sleepiness and Sleep Disturbances Caused by Shift Work." *Journal of the American Medical Association* 313, no. 9 (2015): 961–62.

CHAPTER 8

1. Kouider, S., T. Andrillon, L. S. Barbosa, et al. "Inducing Task-Relevant Responses to Speech in the Sleeping Brain." *Current Biology* 24, no. 18 (2014): 2208–14.
2. Chang, A. M., D. Aeschbach, J. F. Duffy, and C. A. Czeisler. "Evening Use of Light-Emitting eReaders Negatively Affects Sleep, Circadian Timing, and Next-Morning Alertness." *Proceedings of the National Academy of Science USA* 112, no. 4 (2015): 1232–37.
3. Drake, C., T. Roehrs, J. Shambroom, and T. Roth. "Caffeine Effects on Sleep Taken 0, 3, or 6 Hours before Going to Bed." *Journal of Clinical Sleep Medicine* 9, no. 11 (2013): 1195–1200.
4. Afaghi, A., H. O'Connor, and C. M. Chow. "High-Glycemic-Index Carbohydrate Meals Shorten Sleep Onset." *American Journal of Clinical Nutrition* 85, no. 2 (2007): 426–30.
5. Grigsby-Toussaint, D. S., K. N. Turi, M. Krupa, et al. "Sleep Insufficiency and the Natural Environment: Results from the US Behavioral Risk Factor Surveillance System Survey." *Preventive Medicine* 78 (2015): 78–84.
6. Yetish, G., H. Kaplan, M. Gurven, et al. "Natural Sleep and Its Seasonal Variations in Three Pre-Industrial Societies." *Current Biology* 25, no. 21 (2015): 2862–68.
7. Raymann, R. J., D. F. Swaab, and E. J. Van Someren. "Skin Deep: Enhanced Sleep Depth by Cutaneous Temperature Manipulation." *Brain* 131, part 2 (2008): 500–13.

CHAPTER 9

1. Harvey, A. G., and N. Tang. "(Mis)Perception of Sleep in Insomnia: A Puzzle and a Resolution." *Psychological Bulletin* 138, no. 1 (2012): 77–101.
2. Hofer-Tinguely, G., P. Achermann, H. P. Landolt, et al. "Sleep Inertia: Performance Changes after Sleep, Rest and Active Waking." *Cognitive Brain Research* 22, no. 3 (2005): 323–31.
3. Mednick, S., T. Makovski, D. Cai, and Y. Jiang. "Sleep and Rest Facilitate Implicit Memory in a Visual Search Task." *Vision Research* 49, no. 21 (2009): 2557–65.
4. Trauer, J. M., M. Y. Qian, J. S. Doyle, et al. "Cognitive Behavioral Therapy for Chronic Insomnia: A Systematic Review and Meta-Analysis." *Annals of Internal Medicine* 163, no. 3 (2015): 191–204.

CHAPTER 10

1. Van Someren, E. J., C. Cirelli, D. J. Dijk, et al. "Disrupted Sleep: From Molecules to Cognition." *Journal of Neuroscience* 35, no. 14 (2015): 13889–95.
2. Alapin, I., C. S. Fichten, E. Libman, et al. "How Is Good and Poor Sleep in Older Adults and College Students Related to Daytime Sleepiness, Fatigue, and Ability to Concentrate?" *Journal of Psychosomatic Research* 49, no. 5 (2000): 381–90.
3. Morin, C. M. *Insomnia.* New York: Guilford Press, 1996.
4. Thorpy, M., and S. F. Harris. "Can You Die of Insomnia?" [blog post]. *New York Times*, June 24, 2010.

CHAPTER 11

1. Weintraub, K. "Do Sleeping Pills Induce Restorative Sleep?" [blog post]. *New York Times,* December 11, 2015; well.blogs.nytimes.com/2015/12/11/ask-well-do-sleeping-pills-induce-restorative-sleep/?_r=0.
2. Costello, R. B., C. V. Lentino, C. C. Boyd, et al. "The Effectiveness of Melatonin for Promoting Healthy Sleep: A Rapid Evidence Assessment of the Literature." *Nutrition Journal* 13 (2014): 106.
3. Sutton, E. L. "Profile of Suvorexant in the Management of Insomnia." *Drug Design, Development and Therapy* 9 (2015): 6035–42.

CHAPTER 12

1. Chung, S. A., T. K. Wolf, and C. M. Shapiro. "Sleep and Health Consequences of Shift Work in Women." *Journal of Women's Health* 18, no. 7 (2009): 965–77.

CHAPTER 13

1. Riedel, B. W., and K. L. Lichstein. "Insomnia and Daytime Functioning." *Sleep Medicine Reviews* 4, no. 3 (2000): 277–98.
2. Lewith, G. T., A. D. Godfrey, and P. Prescott. "A Single-Blinded, Randomized Pilot Study Evaluating the Aroma of *Lavandula augustifolia* as a Treatment for Mild Insomnia." *Journal of Alternative and Complementary Medicine* 11, no. 4 (2005): 631–37.
3. Lytle, J., C. Mwatha, and K. K. Davis. "Effect of Lavender Aromatherapy on Vital Signs and Perceived Quality of Sleep in the Intermediate Care Unit: A Pilot Study." *American Journal of Critical Care* 23, no. 1 (2014): 24–29.
4. Léger, D., E. Roscoat, V. Bayon, et al. "Short Sleep in Young Adults: Insomnia or Sleep Debt? Prevalence and Clinical Description of Short Sleep in a Representative Sample of 1004 Young Adults from France." *Sleep Medicine* 12, no. 5 (2011): 454–62.
5. Bayon, V., D. Leger, D. Gomez-Merino, et al. "Sleep Debt and Obesity." *Annals of Medicine* 46, no. 5 (2014): 264–72.
6. Sallinen, M., J. Holm, K. Hirvonen, et al. "Recovery of Cognitive Performance from Sleep Debt: Do a Short Rest Pause and a Single Recovery Night Help?" *Chronobiology International* 25, no. 2 (2008): 279–96.
7. Broussard, J. L., K. Wroblewski, J. M. Kilkus, and E. Tasali. "Two Nights of Recovery Sleep Reverses the Effects of Short-term Sleep Restriction on Diabetes Risk." *Diabetes Care* 39 ,no. 3 (2016): 40–41.

CHAPTER 14

1. Honsberg, A. E., R. R. Dodge, M. G. Cline, and S. F. Quan. "Incidence and Remission of Habitual Snoring over a 5- to 6-Year Period." *Chest* 108, no. 3 (1995): 604–9.

CHAPTER 15

1. Aukerman, M. M., D. Aukerman, M. Bayard, et al. "Exercise and Restless Legs Syndrome: A Randomized Controlled Trial." *Journal of the American Board of Family Medicine* 19, no. 5 (2006): 487–93.
2. Marelli, S., A. Galbiati, F. Rinaldi, et al. "Restless Legs Syndrome/Willis Ekbom Disease: New Diagnostic Criteria According to Different Nosology." *Archives Italiennes de Biologie* 153, nos. 2–3 (2015): 184–93.

Index

Suvorexant, 174–75
systemic lupus erythematosus, 21
systemic sclerosis, 21

T
total horseshit (TH), 66
tuberculosis, 153
twilight sleep, *See* paradoxical
    insomnia

U
Uvex glasses, 109, 195

V
valerian, 121, 188
Valium, 171–72
vigilance, 25, 29, 69–70, 74–75, 78–
    80, 135

Z
Zolpidem, 172–74